NEXT GEN PhD

NEXT GEN PhD

A Guide to Career Paths in Science

Melanie V. Sinche

Harvard University Press

Cambridge, Massachusetts
London, England

For Bryan, always

First Harvard University Press paperback edition, 2018
First printing

Library of Congress Cataloging-in-Publication Data

Names: Sinche, Melanie V., author.
Title: Next gen PhD : a guide to career paths in science/
Melanie V. Sinche.
Description: Cambridge, Massachusetts : Harvard University
Press, 2016. | Includes bibliographical references and index.
Identifiers: LCCN 2016015000 | ISBN 9780674504653
(cloth : alk. paper) | ISBN 9780674986794 (pbk.)
Subjects: LCSH: Science—Vocational guidance. | Doctor of
philosophy degree. | Labor supply—Effect of education on.
Classification: LCC Q147 .S55 2016 | DDC 502.3—dc23 LC
record available at https://lccn.loc.gov/2016015000

Contents

Preface   *vii*

Introduction   *1*

Part One: Getting to Know Yourself

1 How to Connect Your Interests to Careers   *9*

2 But I Have No Skills! (and Other Myths)   *21*

3 How to Identify Your Personal Values   *39*

Part Two: Getting to Know the World of Work

4 To Postdoc or Not to Postdoc?   *49*

5 Career Options for PhDs in Science   *79*

6 Strategies for Exploring Careers and
Building Experience   *107*

7 How to Network Effectively   *125*

Part Three: Getting Started on Your Job Search

8 How to Craft Your Individual Development Plan   *145*

9 How to Apply for Jobs   *159*

10 How to Interview and Negotiate   *191*

Conclusion: Making a Successful Transition   215

Appendix A. Identifying Career Pathways for
    PhDs in Science   223

Appendix B. Resources for PhDs   229

Notes   235

References   243

Acknowledgments   251

Index   253

Preface

"I know what my PI does, and I don't want her job . . . but I'm not sure what else is out there."

"I plan to go on the faculty market, but I'm not sure of my chances, so I guess I should have a Plan B."

"I have been on the market for two years now and have not even received an interview. I have always wanted a faculty job, but realize now that it may not happen for me."

For close to twenty years as a career counselor, university administrator, and consultant, I have heard the same refrains. In fact, most of the PhDs and postdocs who sought guidance from me expressed one of the three broad sentiments above.

One of the most distressing stories I encountered came when I was working with a postdoc who had distinguished himself in his field. In fact, the discovery he made was so profound, it appeared not only in top scientific journals but also in mainstream media around the globe. When he came to see me, however, he was utterly crestfallen. He had been on the academic job market for three years and had emerged as the second-choice candidate for a search committee after his final campus visit.

"I can't do this to my family anymore," he shared with me. He had a toddler at the time, and his wife was pregnant with their second child. "I need to move into something permanent. It's not fair to my family." Even this researcher, who had made a major discovery, who had an impressive publication record with first-author papers in top tier journals, and who was coming from one of the most elite institutions on the planet, could not secure a tenure-track position.

As he was leaving my office, he turned and said, "When undergrads in my field ask me about whether they should apply for a PhD program, I tell them not to bother. I encourage them to leave science." I was devastated. If this postdoc from an Ivy League institution, with a major breakthrough in his field to his credit, was unable to move into a faculty position and was in fact *discouraging* young people from entering science, it seems evident to me that the scientific ecosystem in the United States is alarmingly flawed.

Misperception of the Academic Job Market

For decades, senior members of the scientific community have believed the path to success in academic science to be clearly demarcated. Students need simply to enter a prestigious graduate program in an area of interest, take courses, choose a laboratory, work on a research project, defend, and graduate. With luck and perseverance, students would emerge with a publication or two. Following receipt of the doctoral degree, candidates would move directly into a postdoctoral training position, remaining there for a few years to build additional skills, publish more papers, build independence as a researcher, and move into a tenure-track faculty position at the close of the training period.

Unfortunately, the recent state of the academic job market belies the simplicity of this model. Data collected by the National Science Foundation (NSF) on life science PhD recipients show that only 7 percent of candidates hold tenured or tenure-track positions five years after completion of a PhD.[1] Further highlighting the dramatic downturn in hiring is the preponderance of articles in mainstream media describing this phenomenon.[2]

Although careers outside of the academy were once considered "alternative careers," there persists a lively conversation among PhD candidates, postdocs, career counselors, directors, and deans on social media channels and elsewhere that *tenure-track jobs now represent the alternative.*[3] Jobs outside of the tenure track are no longer "alternative," "alt-ac," nor are they "nonacademic," nor "nontraditional." They are, quite simply, career options for PhDs in science.

Indeed, Francis Collins, director of the National Institutes of Health (NIH), recognized the imbalance between the number of doctoral degrees awarded and the number of faculty appointments available. In 2011, he appointed a work group to study the disparity and develop a sustainable model for the growth of the biomedical research workforce in the United States. This study was to include an assessment of present and future workforce needs in a variety of science-based careers. Based on the research conducted by the work group, the NIH created awards entitled "Broadening Experiences in Scientific Training (BEST)," a federally funded program aimed at challenging the way that PhD scientists have been trained in the past, and at improving career readiness for PhD-trained scientists in the United States.

Given the turning tides of the national conversation surrounding PhD training and career development, the time is right for supplying both individual scientists and institutions with a new way of thinking about careers in science. I will use this volume to take science PhDs through the career transition process from start to finish. Through data and illustrations, I will demonstrate how this process has led a multitude of PhDs to satisfying and intellectually challenging careers.

Why This Book?

A new volume on career options for PhDs in science is long overdue. Current offerings are limited in their examination of data on graduate school enrollment as compared to the number of faculty positions available, in their exploration of newly emerging scientific career fields, and in their consideration and application of career development theory to the PhD population. Most are also out of date.

There are some useful books that explore careers for PhDs in *all* disciplines, such as *So What Are You Going to Do with That?* by Susan Basalla and Maggie Debelius, which may be helpful in a general sense to graduate students exploring career options, but it is not focused on the scientific community. Additionally, there are a few titles that focus on the academic job search, such as *The Academic Job Search Handbook* by Julia Miller Vick, Jennifer S. Furlong, and Rosanne Lurie, and *Tomorrow's Professor* by Richard Reis. Both of these volumes are excellent but concentrate primarily on faculty jobs, an occupation that graduate students and postdocs are regularly exposed to throughout their education and training.

Finally, while numerous titles on career choice for the general public are present in the marketplace, PhD scientists deserve a book that focuses on information and options specific to their population.

Why Now?

The discussion of relevant education and training for scientists in the United States and attendant employment outcomes has become omnipresent. Although this conversation tends to focus on systemic issues and national, or even institutional changes, I have endeavored to write a book for the individual scientist. Trying to change the scientific training culture in the United States is akin to moving mountains, so I will instead share what I have learned over time with readers. It is my hope that anyone, anywhere, regardless of institution or pedigree, with a passion for science and questions about science careers, can pick up this volume and learn more about the career development process, either on her own or in groups. I hope that this book and the data contained in it will keep the national conversation going and will contribute to increased satisfaction for individual scientists everywhere.

For Whom?

This book is intended for the over 600,000 students currently engaged in doctoral programs in the United States, as well as the over 60,000 postdoctoral scholars who work alongside them.[4] It is further written for all of the faculty advisors across the country who seek resources on careers outside of their own, and for undergraduates who are undecided about whether to pursue a PhD in science. Indeed, scientists at any level will gain a greater sense of self by completing the assessment exercises outlined in the book, and will learn about the wide range of careers in science that typically require a PhD.

Finally, this book is designed to serve all of the career counselors, directors, and deans of graduate studies and postdoctoral affairs who confirmed for me the need for a comprehensive resource on careers in science for PhDs. As a career center director at a large research university told me, "Graduate students want me to point them to something they can follow."

I was driven to write this book by the thousands of talented scientists I have met throughout my career, and by the stories they have shared. I have met too many scientists who have believed their work to be inferior to others', who have expressed discomfort in admitting their desire to leave the academy, who have been heartbroken by fruitless years on the academic market, and who have at any point felt downtrodden and hopeless.

My message to all of you is twofold: You are not alone, and many others like you have found *satisfying* careers.

Introduction

In the past decade, the number of PhDs conferred throughout the United States in scientific disciplines has risen dramatically, most notably in engineering and biological sciences.[1] There has also been a marked influx of PhD-level scientists trained overseas coming to the United States following graduation. These trends have increased the size of the pool of applicants for tenure-track faculty positions. Yet while the number of potential applicants has grown substantially, the number of tenure-track jobs available has stagnated (Figure I.1).[2]

Compounding the disparity between supply and demand for faculty jobs is the longstanding and persisting culture of the academy, which assumes a natural progression of its graduate students and postdoctoral trainees into tenure-track faculty positions. Confronting these factors may trigger a multi-layered crisis in the life of a PhD-trained scientist. You may feel uncertain about academic job prospects and frustrated in finding careers outside the academy, which receive tepid support from the existing academic culture. You may also feel an acute sense of loss for abandoning the skills and training dedicated solely to this particular path.

Indeed, the current system of scientific training in the United States "has created expectations for academic career advancement that in many—perhaps most—cases cannot be met," said Gregory Petsko, professor of neurology and neuroscience at Weill Cornell

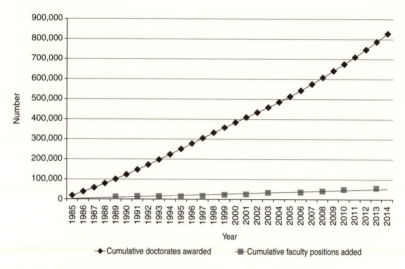

I.1. New faculty positions v. new PhDs

Adapted by permission from Macmillan Publishers Ltd: Maximiliaan Schillebeeckx, Brett Maricque & Cory Lewis, "The Missing Piece to Changing the University Culture," *Nature Biotechnology* 31, no. 10 (October 2013): figure 1, p. 938. © 2013 Nature America, Inc.

Medical College, and chair of the committee that authored the 2014 National Academy of Sciences report, *The Postdoctoral Experience Revisited*.[3]

The Good News

Ongoing debate about how scientists are trained in the United States, and to what end, has been bolstered by national reports that have come to similar conclusions: institutions should, indeed must, introduce graduate students and postdoctoral trainees to a variety of career options earlier in their training.[4]

The primary role of this book, then, is to share the good news that PhDs are gainfully employed in a wide variety of occupations that bring them satisfaction. Don't take my word for it: check out the statistics. In 2013, unemployment among PhDs stood at

2.1 percent, far below the national average.[5] In terms of satisfaction, my own research has demonstrated that PhDs employed both inside and outside of the academy are satisfied with their career choices:

> Of PhDs who graduated between 2004 and 2014 and are currently employed outside of a postdoc position, 80 percent are satisfied with their current occupation.[6]

The secondary role of this book is to provide guidance through the process of developing your career. We will accomplish this goal in three parts. Part One, "Getting to Know Yourself," provides the tools necessary to understand more clearly how your personality, skills, values, and interests can impact your career choice. While PhDs may not be accustomed to this type of self-exploration, it is a necessary step toward preparing to meet the unknown—life after graduate school and postdoctoral training.

"How do I know what my real interests are? How do I find a job that relates to those interests?"   In Chapter 1 we get started with the career-planning process for PhDs that includes self-assessment, career exploration, goal setting, and job searching. Specifically, in this chapter I focus on the connection between interests and career fields. I offer an exercise to assist you in identifying your interests and in channeling them toward activities that appeal to you. Pinpointing your interests in this way will enable you to think more broadly about potential career fields that might be a match.

"But what can I offer an employer? I have no skills!"   We switch gears in Chapter 2 from identifying areas of interest to focusing on areas of strength. Many PhDs seeking careers both within and beyond the academy often assume—incorrectly—that employers will not recognize the skills developed during their graduate and postdoctoral training as useful. Others struggle to acknowledge that they have built any skills at all over the course of their academic training.[7] You may be surprised to learn that the skills sought by a

variety of employers are developed by most PhDs organically over the course of graduate study and postdoctoral training. This chapter includes an exercise to help you identify transferable skills built during your training and connects different skill sets to a variety of occupations.

"How do I find work that is meaningful to me?" Although identifying your interests and skills are worthwhile exercises when considering career options, you will be most successful in finding a robust career match when you also consider your personal values. In Chapter 3 we consider the importance of identifying values in a work setting and demonstrate the process through a written exercise. This chapter concludes with the development of a career profile for you to use as a stepping-stone when you are brainstorming occupations.

In Part Two, we look at employment options for PhDs in science. As competition for faculty positions has grown increasingly fierce, postdoctoral training has become a requirement for securing tenure-track positions and may be required for research positions in different sectors. But postdoctoral training is *not* required for many jobs in science. Chapter 4 answers all your concerns about the postdoctoral appointment, including whether or not you really need to pursue one. It concludes with a discussion of how to make the most of the experience.

"I don't even know what my options are." For those entrenched in the world of academic research, work outside of the university can seem utterly foreign. In Chapter 5, I offer some navigational tools. This chapter also describes a variety of occupations based on interviews with PhD scientists and defines work environments, sample job titles, skills required, and average salary ranges for several career fields.

"Now that I've learned more about my choices, will any of them be a good fit?" In Chapter 6 we explore strategies for learning more

about various career fields, from informational interviewing to job shadowing, interning, and volunteering.

"How can I network comfortably?" Academic researchers often dismiss networking as a negative exercise, requiring "schmoozing." Many believe it is useful only in the business world. However, networking is essential for every job seeker and is key to building relationships in every field. In Chapter 7 I challenge widely held myths about networking and replace them with facts and employment statistics of those who have found success using this job search method. This chapter also offers concrete strategies and tips for building, maintaining, and expanding your network—comfortably—and closes with an outline of several job search methods beyond networking.

In Part Three, we discuss the final stage of career development, the job search itself. First, you must chart a realistic plan for your job search, one that includes some difficult decisions and goal setting. Some of the factors we consider in Chapter 8 include time left to degree, geographic preference, visa status, family concerns, gaps in skills or experiences, and more. Practical decision-making and goal-setting exercises will guide you in this chapter.

"What will I need to apply for this job?" As you think about your job search, you may wonder which documents to prepare for which job applications. The difference between CVs and resumes is outlined in Chapter 9, including what they are, which to choose, how to turn one into the other, and what—and what not—to send with them.

"How do I prepare for an interview?" For those who have gone directly from undergraduate studies into a graduate program, and then perhaps on to a postdoc, interviewing and negotiating for a position may seem overwhelming. We fill in the gaps in Chapter 10 by describing the interview process and what types of interviews

you may encounter. Tools, tips, salary guidelines, and sample state-ments and questions to practice will prepare you for both the inter-view and subsequent salary negotiations.

"What do I need to know on my first day of work?" To make a successful transition to the working world, you'll need to know something about the culture of work. We conclude by looking at professional etiquette and describing accepted norms and practices in the United States. This final chapter also suggests several strate-gies for making the most of a new position, including setting goals with the new employer, keeping the lines of communication open, and more.

PART ONE

Getting to Know Yourself

CHAPTER 1

How to Connect Your Interests to Careers

Congratulations! You are either engaged in or have completed a PhD program, an incredible achievement. And now for a reality check: You are not your PhD. You are a culmination of all that has come before in your life, all that happened during, and all that has happened since you earned your PhD. You are the child who was fascinated by science in your early school years. You are the high school student who participated in science fairs. You are the college student leader. Graduate peer mentor. Rock climber. Tutor. Artist. Writer. Member of a particular culture. Citizen of a particular country. Sibling, child, parent, aunt, uncle, friend, mentor, teacher, researcher. Given the rich fabric of your history and the multiple roles you currently play, you will need time and space to reflect on who you are and what is most important to you before launching into a job search.

> *Q:* The thought of self-exploration makes me totally uncomfortable. Is introspection really necessary for effective career planning?
>
> *A:* Yes! If you spend time on self-reflection and assessment, it is more likely that you will land in a career that is satisfying, where you can be productive, and in which you may remain for a substantial period of time.[1]

Acknowledging Your Thoughts about the Future

When you think about your future after your PhD or postdoc, what feelings can you identify? Anxiety? Apprehension? Do you find the whole idea overwhelming? Do you prefer to put your head down and focus on your research, rather than focus on yourself and your future?

If so, you are not alone. These thoughts and feelings are common among PhDs completing their graduate or postdoctoral training and considering their next steps. I know this because I have worked with thousands of PhDs and postdocs over the years who have come to see me for career advice with no small amount of trepidation. Despite their initial pessimism, most of these trainees do at last recognize the value of introspection in getting to an answer about what they might like to do for the rest of their lives.

I once worked with a postdoc who came to see me with this very question in mind, but was skeptical when I explained the importance of self-assessment in finding a job that would be a good fit for him. He was resistant at first, but the more we talked, the more he came to recognize that the thoughts he was sharing with me as he reflected on himself and his interests were indeed pointing to a particular career—a career in science writing. His face lit up every time he mentioned writing about his own work, and when he considered writing about the work of other scientists for a lay audience.

We met over several months, going through practical exercises like those you will find in this book, and we synthesized his responses not only to validate that science writing might be a suitable choice for him, but also to compile potential interview responses, for the anecdotes he shared demonstrated his evident love of writing and enthusiasm for a range of scientific disciplines.

Stages of Career Development

When nearing the end of a degree program, postdoc appointment, or job, most people tend to jump directly into the fourth or final stage of career development, the job search itself. This is perfectly

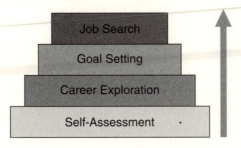

1.1. The four stages of career development

understandable, since most of us need a job to eat, pay our bills, cover our rent or mortgage, and clothe ourselves or our children. Some send resumes blindly and hope someone notices their work and calls for an interview. This job search method, while easy and free, can be time-consuming, or worse, utterly disheartening, because this method rarely results in an interview. However, the more time you spend on self-assessment, which is the foundational or first stage of career development, the greater the chance you will identify an occupation or two that are an appropriate fit for you. You can then go on to explore those careers in detail, network, apply for jobs, and ultimately secure a position.

In career development, we can identify four distinct stages: self-assessment, career exploration, goal setting, and job searching (Figure 1.1). Self-assessment is the first, and arguably the most critical, step in the career-planning process. Once you know more about yourself and what is most important to you, you will be better positioned to move into the second stage, career exploration. In this stage, you will be delving into the specifics of what a job entails, what skills are required to enter a particular field, and how to make yourself more attractive for those kinds of positions. Goal setting follows career exploration. You will set concrete goals to fill any gaps in your skill set in order to be more competitive for a particular career. The fourth and final stage is the job search process, when you will prepare your materials, network, apply for jobs, interview, and negotiate offers.

What Does Self-Assessment Entail?

Self-assessment is a process that involves exploring your personality, your interests, your skills, and your values. In this and the following two chapters, you will develop a greater understanding of what interests you, what you are good at, and what you care about most.

Self-assessment can be a daunting task for some, perhaps even more so for scientists, because they are committed to logic and assessing data objectively. The other three stages of career development are vastly more concrete, requiring you to use both your well-honed research skills (to explore or apply for jobs) and your organizational skills (mapping out your goals and documenting your progress).

So why engage in this activity? Take a look at Figure 1.1 for motivation. The very structure of this figure suggests that the more time you spend in the first stage, that is in the *foundation,* the more firm your own foundation will be.

> Q: Are there any risks in conducting a self-assessment, other than my own discomfort?
>
> A: Yes—it is important to remember that self-assessment is a process that takes time and effort, and that it is not prescriptive. There is no right answer that any assessment tool or person can give you about the best career choice for you. It takes a significant amount of time and reflection to identify occupations of interest and to make an informed choice about a career.

This chapter contains exercises to help you identify your interests and find your ideal job. In the two chapters that follow, you will determine which skills you would like to use in your next position and engage in a deeper exploration of your personal values. Taken together, these three chapters will give you a better sense of who you are and enable you to assess different occupations and their potential fit.

Exercise: Identify Your Interests

Now that you recognize the import of exploring yourself and your interests, how might you actually go about doing it? To get you started, here are a few ways you can identify your interests.

Self-Reflection

Take a moment to think carefully about topics you find compelling. How do you spend your free time? Which books do you read? Which shows do you watch? Which blogs do you follow? These types of questions might help you identify areas of interest as you begin your exploration.

It will help you to go through this process methodically by journaling in a career notebook or by keeping your responses in a file in your laptop. What themes emerge? For example, if you are constantly reading political blogs or checking in on web discussions on politics throughout the day, how might you weave that interest into a potential career? Might you consider the AAAS Science and Technology Policy Fellowship, where you might be placed in Congress, working closely with a legislator to translate complex scientific ideas? Document your interests when you have a few quiet moments in the lab or office. Jot down ideas as they occur, perhaps after reading a journal article or news feed about a career you find compelling enough to research further.

Feedback from Others

You can—and should—talk with friends, family, partners, colleagues, those who know you best, for help in determining your interests. These people may be members of your research group or collaborators, if you feel comfortable discussing yourself openly with them. Ask the folks in these groups the following:

- What topics do you tend to focus on?
- When are you most animated?
- What do you seem most excited about?

Chances are good that people who know you well will be able to point to one or several themes that excite you—maybe grant writing, presenting your research, or mentoring others, for example, or your subject area, or maybe a topic outside of science.

Feedback from Career Counselors and Mentors

Check with your institution to find out whether there are career counselors who focus on serving PhDs.[2] In most cases, you will find these people in career centers, institutional postdoc offices, or perhaps in human resources units. Career counselors are highly skilled and trained to serve people like you—highly educated scientists who are considering their next steps. Do not miss the opportunity to work with this valuable group of experts as you continue on your path toward self-knowledge. They most likely will offer individual career counseling sessions, which have the added benefit of confidentiality, in addition to group workshops for PhDs in science.

Mentors can support you through the process of self-assessment and, in some cases, share their own experiences of self-assessment with you. This group may be comprised of one or more of the following: your PhD advisor, your postdoc advisor, training directors at your institution, counselors or directors within career centers, postdoc offices, scientists you've met at conferences, collaborators, or even peers. Try meeting over lunch or coffee to discuss your career ideas with your mentors. Here again, as with family and friends, these important players in your work and life may be able to identify interests that you mention through the course of your conversations and interactions with them.

Peer mentorship is becoming more and more prevalent in the scientific community. For example, when I worked at a large public research university, I hosted a small group of PhDs and postdocs who met weekly over lunch to discuss career plans and ideas. Individuals took turns describing their interests to one another, and the remainder of the group brainstormed career options for them. Consider forming a peer mentorship lunch group to discuss career ideas and options.

Exercise: Find Your Ideal Job

There are several resources that exist both online and in print to help you identify your interests in a more tangible way than through open-ended discussions: here is one of my favorites. As you progress through this activity, be sure to write down whatever words or phrases come to mind. This is not the time to censor yourself—this is the time to be open to responses that enter your mind immediately, unbiased and unmediated.

Use your career notebook or laptop to jot down your answers to the following:

General Interests
1. What have you studied in the past? (BS, MS, PhD, other degrees and disciplines)
2. Why did you choose these fields?
3. What topics are you most passionate about? Think not only about your work and educational background, but also about your leisure activities, and social outlets.

Activities
4. What activities inspired you as an undergraduate? As a graduate student? Postdoc?
5. As you think about your current work, which activities do you enjoy most?
6. As you think about your current work, which activities do you enjoy least?

Interaction with Science
7. In what way would you like to interact with science in your next job? Would you like to be exposed to science

broadly? Or work on one particular research problem in a narrowly defined discipline? Mark this point on the line in Figure 1.2.

| Broad exposure to science | Deep and narrow interaction with science |

1.2. Choose your preferred level of interaction with science in your next job

Environment
   8. Where would you like to live?
   9. What size organization would you like to work in?
   10. What type of environment would you like to work in? Corporate? In a less formal office setting? Outside? In a laboratory?

People
   11. Describe the people you enjoy being around.
   12. Would you prefer to work alone, in a group, or have occasional contact with others?

Draft Your Ideal Job Description
Read through your responses to the questions above and try to identify patterns. Synthesize your answers and draft your ideal job description.

Answering these questions and creating a draft of what to look for in a job is useful for getting conversations started during career counseling appointments. You can also take your work to a friend, colleague, or family member and discuss them in more detail. Your answers to these questions and the subsequent conversations they stimulate very often will help you to get past your preconceived ideas about your future. Then you'll be able to move forward and compile a list of potential areas to explore.

Link Your Leisure Interests to Jobs

While working at a research university, I met a postdoc who was just starting the career development process and did not know where to begin. We had a discussion around the questions above, and through that conversation, I came to learn about her interest in volunteering to spend time with kids in science. She had volunteered for several science festivals, offered to run experiments for kids, and had thoroughly enjoyed those interactions—more so even than the work she was doing at the time in her laboratory. As a result of that conversation, she went on to explore careers in science education and outreach while continuing to pursue this passion outside of work. She contacted many professionals in the field she was investigating and ended up having helpful conversations with those scientists—learning more about career options in the process. Self-assessment ultimately helped her to develop a better understanding of her own interests.

Formal Assessments

Another way to assess your interests is to take part in formal interest assessments. Some of these are available online, some are free, and some require a fee for use. To find out which assessments your institution uses, visit your career center, postdoc office, or human resources website.

While these may be worthwhile tools to use in the assessment process, it is critical to recognize their limitations. These inventories will not present you with all the answers to your career questions. Greater and varied introspection is required, although these tools can be used as part of that process.

Career Beliefs Inventory

The Career Beliefs Inventory (CBI) is a relatively inexpensive assessment you can take online. This assessment, developed by John

Krumboltz, measures career readiness and may be a good fit for you if you are trying to identify obstacles to making a career decision. This inventory looks at your assumptions and beliefs and attempts to clarify any that might be preventing you from reaching your career goals.

Strong Interest Inventory

The Strong Interest Inventory (Strong) is the oldest interest assessment in the United States. Published in 1927 by a military psychologist, E. K. Strong Jr., and updated six times over the years to reflect the changing times, the Strong offers users an opportunity to consider their interest areas through a checklist of occupations, leisure interests, and coursework. The Strong can be valuable in generating ideas for exploration, as each user will emerge with a list of basic interest areas and occupations to explore. However, it is important to note that the Strong is not prescriptive—it is simply another tool that job seekers can use to investigate different fields. Additionally, the Strong cannot be used alone, but must be interpreted by a qualified professional, such as a career counselor, trainer, or therapist.

Myers-Briggs Type Indicator

The Myers-Briggs Type Indicator (MBTI) was first published in 1943.[3] This instrument is designed to help users understand the ways in which they see the world, take in information, and make decisions. Knowing more about yourself and your personality can also be useful in making a determination about whether a particular career is a good fit for you. As with the Strong, the MBTI must be administered and interpreted by a trained professional. The MBTI is the world's most widely used personality assessment, and many attempts have been made to replicate the MBTI and offer it free of charge on the Internet. Be aware that the only MBTI instrument that has been tested for reliability and validity is the one published by CPP, Inc.

myIDP

Developed in 2011, myIDP is another online tool that PhDs can use in the self-assessment process. This assessment is free and open to all. The exercises on the myIDP website include a list of potential interest areas, skills, and values that users can complete on their own. This site also includes a wealth of information on a wide variety of careers in science, and can be used for further career exploration. Finally, it includes a goal-setting feature with automated email reminders, based on your preferences.[4]

Which formal assessment should you choose, and how should you use it? Some are self-directed, and others require the assistance of a trained career counselor for interpretation and guidance. If you are feeling overwhelmed and have no ideas about where to start with this process, it might be worthwhile for you to take a formal assessment like the Strong Interest Inventory. I like to use this tool to assist PhDs in generating career ideas when they are having trouble articulating their interests and expressing them to me.

Try meeting first with a career counselor or advisor to discuss options. Ideas may arise from your discussion, or you may be offered the opportunity to engage in more formal exercises like those listed above. If you don't have access to a career counselor at your institution, you might look to the National Board for Certified Counselors to find someone who can guide you through this process.[5]

CHAPTER 2

But I Have No Skills! (and Other Myths)

"But I have no employable skills!" I have heard this statement and others like it repeatedly from PhDs in science over the past decade. "Why would anyone ever want to hire me?" The truth is that candidates who have successfully completed a PhD in science have myriad skills of value to employers of all kinds, from universities to research institutes, government agencies to nonprofits, start-ups to professional societies. Identifying what those skills are, as well as providing evidence that you have used those skills, are the goals of this chapter.

By the time you emerge from a doctoral program or postdoctoral training, you will have developed discipline-specific knowledge. That toolkit is easily recognizable by others, though sometimes PhDs doubt their expertise in their own research area. Many PhDs feel that they are impostors and fear that someone, somewhere will discover that they are frauds. In her book *The Secret Thoughts of Successful Women*, Valerie Young writes about how many bright and capable men and women still doubt their own work and worth.[1]

This scenario is quite common. I experienced it as a graduate student myself. During both my coursework and my thesis-writing stages, I was plagued by thoughts of how much everyone *else* knew, how little I knew, and how I was fooling everyone and did not deserve to be in my graduate program. There were so many talented graduate students around me with more to offer; I felt that I did not belong there. These thoughts were powerful, but not necessarily rational. If you are experiencing similar thoughts and doubt yourself and your abilities, remember that the graduate admissions committee, your PhD advisory committee, your faculty mentor, and others most likely believe in you and your work, and they want you to succeed. For more support and tools to manage irrational thoughts and feelings, you might consider visiting the counseling center at your institution, or seeing a therapist with experience working with PhDs.

Beyond the impostor syndrome, there are a few other statements you may have heard.

Myth: I am trained for only one job.

Reality: While PhDs have historically trained for jobs in the professoriate, the vast majority of recent PhD grads are employed outside of faculty ranks. Graduate and postdoctoral training require you to develop a broad set of skills that are in demand across a wide range of career paths. These skills include the ability to find information, analyze data, and solve problems. They are of enormous value to employers both within and beyond the academy. See the career options listed in Chapter 5 for more evidence of job choices for PhDs.

Myth: My research is so specialized that there are very few applications for it.

Reality: Even if you have investigated a narrow slice of a broader field, you have developed skills over the course of your training that are transferable to many different occupations. Regardless of the specialization of your work, for example, you have acquired skills that are important for success in a broad range of career fields.

Myth: But I have no skills!

Reality: Many PhDs I have worked with have expressed this feeling, and have had trouble identifying skills they possessed.

I once worked closely with a postdoc in evolutionary biology, for example, who shared this sentiment in a career counseling session. He was convinced that he did not have any employable skills. As we began talking through his research project, however, it became clear that he was a critical member of his research group. He approached every potential problem with a uniquely creative point of view. Over time, he came to realize that he could parlay his creativity into a career developing science outreach programs for the public through a nonprofit organization. He is now happily employed and making contributions to his new team's daily work.

Like him, you actually *do* have skills, though it may take time to identify them, embrace them, and articulate their usefulness to others. We will begin to do that work in this chapter.

Skills Developed by PhDs

For the past decade, professional societies, individual researchers, graduate programs, university administrators, career counselors, and other groups have conducted research to identify skill sets developed by PhDs in science. In 2009, for example, the National Postdoctoral Association (NPA) compiled six core competencies that lead to a scholar achieving "intellectual and professional independence and success."[2] The authors of the career development website myIDP referred to the NPA's list as they developed the skills assessment section of their site.[3] Similarly, Michigan State University's Career Success program outlines six skills areas that employers most frequently cite as important when they are looking at job candidates with PhDs. Michigan State's program includes a free assessment tool for PhDs to gauge how confident they are in each area.[4]

Using this work as a baseline, I conducted survey research in 2015 to determine which skills recent PhDs had developed organically in the course of a doctoral program, and which of these skills were important for success in their current jobs. The list below includes responses from 3,816 PhDs to the statement: "I developed/continued to develop this skill during my doctoral program." Respondents were asked to rate their agreement on a Likert-type scale from "Strongly Disagree" to "Strongly Agree," and those percentages on the left reflect those who selected "Agree" or "Strongly Agree" in response to each proposed skill. The column on the right lists the percentages of those who rated these skills as "Very Important" or "Extremely Important" in answer to the question: "Which of the following skills are important for success in your current position?"

Skills Acquired during Training	Skill	Skills Important for Success on the Job
95%	Discipline-specific knowledge	80%
95%	Ability to gather and interpret information	92%
93%	Ability to analyze data	83%
83%	Ability to make decisions and solve problems	93%
82%	Oral communication skills	93%
82%	Written communication skills	91%
71%	Ability to learn quickly	89%
67%	Creativity/innovative thinking	82%
66%	Ability to manage a project	87%

As you can see, this research suggests that multiple skills are developed organically through doctoral training, and that these same skills are indeed important for success in a variety of career fields. Furthermore, the self-assessment of skills demonstrated by former PhDs is quite positive in terms of outcomes recognized after completing graduate training.

Transferable Skills: Meaning and Context

Now that we have identified skills PhDs tend to develop organically during their graduate work, let's explore the process of translating skills that are transferable to different jobs and contexts. For example, if you have acquired strong project management skills through your thesis or postdoctoral research, how might those skills transfer to a position in consulting? How would you describe them to a potential employer? Job descriptions for associates in consulting frequently list project management as a skill required for success on the job. Take what you have learned by managing your dissertation project and reframe that skill set in a way that would be meaningful to a consulting firm, using language found in a sample job description.

Building this kind of evidence of skill development and application is critical for every job candidate, and perhaps more so for PhDs considering jobs outside of the academy. You need to be explicit in your description of your work to demonstrate that you indeed possess the skills required by the employer.

How can you emphasize your transferable skills most effectively? Take a look at this next exercise and try to identify the transferable skills necessary for success.

Exercise: Transferable Skills

Ellen Elliott, a postdoc interested in entering the field of patent law, is applying for a job. Read through the job description and her sample CV that follows, scanning the job description at least twice and noting skills required for this job.

A nationally prominent law firm is seeking a Patent Agent to join its Intellectual Property (IP) Practice group in its downtown Washington, DC office. The Patent Agent's duties will include conducting analysis and interpretation of patent searches; preparing written opinions on issues of patentability; preparing and filing patent applications and oppositions; and performing other duties as assigned. The successful candidate will have an MS, PhD, or equivalent technical experience in any branch of Biology, Chemistry, or Biochemistry. Excellent analytical, interpersonal, and communication skills are a must for this position. The candidate should also have strong project and client management skills.

SAMPLE CV BEFORE ANALYSIS

ELLEN NICHOLE ELLIOTT
65 Penn Drive
West Hartford, CT 06119
ellen.elliott@jax.org

Academic Training

Present:	Postdoctoral Associate, The Jackson Laboratory for Genomic Medicine. Advisor: Dr. Adam Williams, Ph.D.
August 2015:	Ph.D., University of Pennsylvania Perelman School of Medicine, Cell and Molecular Biology Program, Philadelphia, PA. Advisor: Dr. Klaus H. Kaestner, Ph.D.
May 2010:	Indiana University, Bloomington, IN. Biology B.S., Chemistry minor, Cumulative GPA: 4.0.
May 2010:	Indiana University, Bloomington, IN. Neuroscience B.S., Psychology minor, Cumulative GPA: 4.0.

Laboratory Experience

Date:	August 2015–Present
Position:	Postdoctoral Associate
Mentor:	Dr. Adam Williams, Ph.D.
Institution:	The Jackson Laboratory for Genomic Medicine
Description:	Analysis of long noncoding RNA (lncRNA) expression and function in T helper type 2 cells, with a focus on the role of lncRNAs in asthma.

Date:	May 2011–August 2015
Position:	Graduate Student
Mentor:	Dr. Klaus H. Kaestner, Ph.D.
Institution:	Institute for Diabetes, Obesity and Metabolism, Perelman School of Medicine, University of Pennsylvania, Philadelphia, PA

Description: Focused on the function of DNA methyltransferases during intestinal epithelial development. Discovered that the maintenance DNA methyltransferase *Dnmt1* critical to maintain intestinal progenitor cells in the perinatal period.

Techniques: Mouse genetics and dissection, RNA-Seq, Bisulfite-sequencing, intestinal organoid culture, immunohisto-chemistry staining and analysis, microsatellite instability assays

Date: September 2006–June 2010
Position: Undergraduate Research
Mentor: Dr. Yves V. Brun, Ph.D.
Institution: Department of Biology, Indiana University, Bloomington, IN
Description: Transoposon mutagenesis to discover genes required for surface adhesion and motility in the bacterium *Asticca-caulis biprosthecum*. Helped with the discovery of a holdfast-shedding mutant further characterized in the related bacterium *Caulobacter Crescentus*.
Techniques: Transposon mutagenesis, biofilm adhesion and swarming motility assay, bacterial cloning, lectin binding and fluorescent microscopy

Date: Spring 2009
Position: Undergraduate Research Assistant
Mentor: Dr. Preston E. Garraghty, Ph.D.
Institution: Department of Psychological and Brain Sciences, Indiana University, Bloomington, IN
Description: Assisted with project analyzing effects of operant conditioning on rat somatosensory cortex.
Techniques: Microscopic neuronal tracing and analysis, Morris water maze and Skinner box learning paradigms

Date: Summer 2006
Position: Undergraduate Research Assistant as part of Integrated Freshman Learning Experience (IFLE) Lab Research, Medical Sciences Program
Mentor: Dr. Claire E. Walczak, Ph.D.
Institution: Medical Sciences Program, Indiana University, Bloomington, IN

Description:	Assess localization of the protein MCAK (mitotic centromere-associated kinesin) during mitosis.
Techniques:	Antibody staining and fluorescent microscopy of mitotic *Xenopus* oocyte extracts

Grant Support

Fall 2015–Fall 2017:	Pyewacket Fellowship, The Jackson Laboratory
July 2012–April 2014:	University of Pennsylvania Developmental Training Biology Grant, NIH T32-HD07516-15
Fall 2009–Spring 2010:	Barry Goldwater Scholarship and Excellence in Education Program
June 2008–June 2009:	Beckman Scholars Program
Spring 2007:	HHMI Capstone Undergraduate Research Training Grant
Fall 2007–Spring 2010:	STARS (Science, Technology, and Research Scholars) undergraduate research program at Indiana University

Peer Reviewed Publications

Sheaffer K.L.*, **Elliott E.N.***, Kaestner K.H. DNA hypomethylation contributes to genomic instability and intestinal cancer initiation. In press, *Cancer Prevention Research*, PMID: 26883721.

Elliott E.N.*, Sheaffer K.L.*, Kaestner K.H. The 'de novo' DNA methyl-transferase Dnmt3b compensates the Dnmt1-deficient intestinal epithelium. *eLife* 2016;10.7554/eLife.12975

Elliott E.N. and Kaestner K.H. Epigenetic regulation of the intestinal epithelium. *Cell and Molecular Life Sciences* 2015. 72: 4139–4156.

Elliott E.N., Sheaffer K.L., Schug J., Stappenbeck T.S., Kaestner K.H. Dnmt1 is essential to maintain progenitors in the perinatal intestinal epithelium. *Development* 2015. 142: 2163–2172.

Sheaffer K.L., Kim R., Aoki R., **Elliott E.N.**, Schug J., Burger L., Schübeler D., Kaestner K.H. DNA methylation is required for the control of stem cell differentiation in the small intestine. *Genes & Development* 2014. 28: 652–664.

* These two authors contributed equally to this study.

Zhe Wan, Pamela J. Brown, **Ellen N. Elliott**, Yves V. Brun. The adhesive and cohesive properties of a bacterial polysaccharide adhesin are modulated by a deacetylase. *Molecular Microbiology* 2013. 88: 486–500.

Seminars

Ellen N. Elliott and Klaus H. Kaestner. Dnmt1 is essential to maintain progenitors in the perinatal intestinal epithelium. University of Pennsylvania, Perelman School of Medicine, Intestinal Stem Cell Club Monthly Meeting. Philadelphia, PA, September 2014.

Ellen N. Elliott and Klaus H. Kaestner. Dnmt1 is essential to maintain progenitors in the perinatal intestinal epithelium. University of Pennsylvania, Perelman School of Medicine, DSRB Graduate Program Seminar. Philadelphia, PA, September 2014.

Ellen N. Elliott and Klaus H. Kaestner. The role of the DNA methyltransferase Dnmt1 in intestinal epithelial development. Visiting graduate student seminar, Dr. Thaddeus Stappenbeck Lab, Washington University, St. Louis, MO, November 2013.

Ellen N. Elliott and Klaus H. Kaestner. The role of the DNA methyltransferase Dnmt1 during intestinal epithelial development. University of Pennsylvania, Perelman School of Medicine, Developmental Biology Training Grant Mini-Retreat, Philadelphia, PA, June 2013.

Ellen N. Elliott and Klaus H. Kaestner. The role of the DNA methyltransferase Dnmt1 during intestinal epithelial development. University of Pennsylvania, Perelman School of Medicine, DSRB Graduate Program Seminar. Philadelphia, PA, May 2013.

Ellen N. Elliott and Klaus H. Kaestner. The role of the DNA methyltransferase Dnmt1 during intestinal epithelial development. University of Pennsylvania, Perelman School of Medicine, Intestinal Stem Cell Consortium Monthly Meeting. Philadelphia, PA, November 2012.

Selected Posters

Ellen N. Elliott, Karyn L. Sheaffer, Jonathan Schug, Thaddeus S. Stappenbeck, Klaus H. Kaestner. Dnmt1 is essential to maintain progenitors in the perinatal intestinal epithelium. Keystone Symposia on Molecular and Cellular Biology, DNA Methylation and Epigenomics, Keystone, CO, March 2015.

Ellen N. Elliott, Karyn L. Sheaffer, Jonathan Schug, Thaddeus S. Stappenbeck, Klaus H. Kaestner. Dnmt1 is essential to maintain progenitors in the perinatal intestinal epithelium. Perelman School of Medicine Cell and Molecular Biology Graduate Program Symposium. Philadelphia, PA, October 2014.

Ellen N. Elliott, Karyn L. Sheaffer, Klaus H. Kaestner. The role of the DNA methyltransferase Dnmt1 during intestinal epithelial development. Perelman School of Medicine Cell and Molecular Biology Graduate Program Symposium. Philadelphia, PA, September 2013.

Ellen N. Elliott, Karyn L. Sheaffer, Klaus H. Kaestner. The role of the DNA methyltransferase Dnmt1 during intestinal epithelial development. FASEB Science Research Conference: Gastrointestinal Tract XV: Epithelia, Microbes, Inflammation and Cancer. Steamboat Springs, CO, August 2013.

Ellen N. Elliott, Karyn L. Sheaffer, Klaus H. Kaestner. The role of the DNA methyltransferase Dnmt1 during intestinal epithelial development. The NIH Center for Molecular Studies in Digestive and Liver Diseases and The Penn/CHOP Center for Digestive, Liver & Pancreatic Medicine Division of Gastroenterology 14th Annual Symposium. Philadelphia, PA, June 2013.

Honors and Awards

Spring 2010:   Indiana University Department of Biology Outstanding Honor's Thesis Award

Spring 2010:   Indiana University Provost's Award for Undergraduate Research and Creative Activity

Fall 2009:   Phi Beta Kappa

Spring 2008:   Fox Glenn Research and Education Fund, Indiana University Biology Department

Spring 2008:   Fernandus and Elizabeth Payne Scholarship, Indiana University Biology Department

Teaching Experience

The Jackson Laboratory for Genomic Medicine, Williams Lab, winter/spring 2016. Mentored and trained Middlesex Community College Intern on CRISPR-Cas9 molecular cloning, RNA and DNA purification techniques, and qRT-PCR.

University of Pennsylvania Perelman School of Medicine, Department of Genetics, Kaestner Lab, summer 2013. Mentored and trained high school student on PCR, staining, and microscopy techniques.

University of Pennsylvania Perelman School of Medicine, Biomedical Graduate Studies Program Teaching Assistant Spring 2013 for BIOM555: Control of Gene Expression. Responsibilities included grading exams and midterm papers, leading weekly discussion section, attending biweekly lectures.

University of Pennsylvania Perelman School of Medicine, Gastroenterology Dept. Undergraduate Student Scholars Program, summer 2012. Trained and mentored undergraduate in Kaestner lab on PCR, staining, and microscopy techniques, and oral presentation of research.

Indiana University Research Experience for Undergraduates (REU) Program, Bloomington, IN, summer 2009. Mentored and trained undergraduate in Brun lab to perform transposon mutant characterization.

Administration

Kaestner Lab website manager, 2013–2015. I was trained on Adobe Contribute software, which allowed me to update information on current lab members, alumni, publications. I was also able to add or change images to represent Kaestner lab research expertise.

University of Pennsylvania Perelman School of Medicine, Developmental Stem Cell and Regenerative Biology Program Student-run journal club, 2014–2015 academic calendar year. I organized and scheduled rooms, recruited volunteers to present papers, and delegated people to bring food and drinks, in addition to actively participating bi-monthly in the journal club.

References

Available upon request.

1. List the skills you observed in the job description in your career notebook or on your laptop.
2. Read through Elliott's CV. Would you call this person for an interview? As you assess her CV, consider how easy or difficult it is to find evidence to support her candidacy for the job. Are you able to find skill sets that could be transferred to this type of work?

Highlight Transferable Skills

To make her CV stronger for this position, Elliott might consider highlighting her transferable skills by adding a Summary of Qualifications. This can be an effective tool in framing a CV for the reader. Here is a summary we might consider for Elliott's CV:

PhD-level biologist with a proven track record of delivering high-quality, scientific results through critical analysis and interpretation. Exceptional written and oral communication skills. Committed to applying scientific and analytical skills to assist in protecting inventions.

By using a summary, she highlights the particular skill set sought by the law firm in Washington, as her updated CV shows.

REVISED CV WITH A SUMMARY OF QUALIFICATIONS

ELLEN NICHOLE ELLIOTT
65 Penn Drive
West Hartford, CT 06119
ellen.elliott@jax.org

Summary of Qualifications

- PhD-level biologist with a proven track record of delivering high-quality, scientific results through critical analysis and interpretation
- Exceptional written and oral communication skills
- Committed to applying scientific and analytical skills to assist in protecting inventions

Academic Training

Postdoctoral Associate   August 2015–present
The Jackson Laboratory for Genomic Medicine, Farmington, CT
Advisor: Dr. Adam Williams, Ph.D.

Ph.D., Cell and Molecular Biology   August 2015
University of Pennsylvania Perelman School of Medicine,
Philadelphia, PA
Advisor: Dr. Klaus H. Kaestner, Ph.D.

B.S., Biology, Chemistry minor   May 2010
B.S., Neuroscience, Psychology minor
Indiana University, Bloomington, IN
Cumulative GPA: 4.0.

Scientific Communication Skills

- Served as first author on four peer-reviewed publications
- Invited speaker to seminars and national conferences, delivering oral presentations as well as poster presentations
- Served as teacher, trainer, and mentor for undergraduate students in science
- Reviewed research abstracts and applications for granting agencies and student funding mechanisms

Honors and Awards

Indiana University Department of Biology Outstanding Honor's Thesis Award	Spring 2010
Indiana University Provost's Award for Undergraduate Research and Creative Activity	Spring 2010
Phi Beta Kappa	Fall 2009
Fox Glenn Research and Education Fund, Indiana University Biology Department	Spring 2008
Fernandus and Elizabeth Payne Scholarship, Indiana University Biology Department	Spring 2008

Communications Experience

Blogger, The Jackson Laboratory for Genomic Medicine, Farmington, CT   Spring 2016
- Authored brief articles on new techniques and advances in science
- Compiled information and wrote blog posts to capture information about laboratory events

Journal Club Co-Leader, The Jackson Laboratory for Genomic Medicine, Farmington, CT Summer 2016
- Led group of 14 undergraduates with a co-facilitator in weekly study of scientific journal articles
- Transformed complex scientific concepts into clear, accessible language for the students
- Assisted in teaching group to dissect and elucidate scientific papers for a variety of audiences

Website Manager, University of Pennsylvania Perelman School of Medicine, Department of Genetics, Kaestner Lab, Philadelphia, PA 2013–2015
- Composed profiles of current lab members, alumni, and an overview of relevant publications using Adobe Contribute software
- Updated images to represent novel research in the Kaestner lab

Student Leader, Journal Club, University of Pennsylvania Perelman School of Medicine, Developmental Stem Cell and Regenerative Biology Program 2014–2015
- Facilitated learning for group of students through bi-monthly of journal club
- Recruited volunteers to present papers

- Actively participated in discussions
- Delegated responsibilities for club to various members
- Organized and scheduled rooms

<u>**Peer Reviewed Publications**</u>

Sheaffer K.L.*, **Elliott E.N.***, Kaestner K.H. DNA hypomethylation contributes to genomic instability and intestinal cancer initiation. In press, *Cancer Prevention Research*, PMID: 26883721.

Elliott E.N.*, Sheaffer K.L.*, Kaestner K.H. The 'de novo' DNA methyl-transferase Dnmt3b compensates the Dnmt1-deficient intestinal epithelium. *eLife* 2016;10.7554/eLife.12975

Elliott E.N. and Kaestner K.H. Epigenetic regulation of the intestinal epithelium. *Cell and Molecular Life Sciences* 2015. 72: 4139–4156.

Elliott E.N., Sheaffer K.L., Schug J., Stappenbeck T.S., Kaestner K.H. Dnmt1 is essential to maintain progenitors in the perinatal intestinal epithelium. *Development* 2015. 142: 2163–2172.

Sheaffer K.L., Kim R., Aoki R., **Elliott E.N.**, Schug J., Burger L., Schübeler D., Kaestner K.H. DNA methylation is required for the control of stem cell differentiation in the small intestine. *Genes & Development* 2014. 28: 652–664.

Zhe Wan, Pamela J. Brown, **Ellen N. Elliott**, Yves V. Brun. The adhesive and cohesive properties of a bacterial polysaccharide adhesin are modulated by a deacetylase. *Molecular Microbiology* 2013. 88: 486–500.

* These two authors contributed equally to this study.

Identifying Your Particular Skill Set

Now that we have transformed a CV to highlight transferable skills sought by an employer through a job ad, it is time to identify the particular skills *you* bring to the table, along with evidence demonstrating those skills. What professional assets have you already developed?

Exercise: Skills Identification

Think about the day-to-day, week-to-week, or month-to-month activities of your graduate program or postdoc training. Consider activities you engage in or have engaged in both inside and outside of your discipline, program, or research project.

1. List every major activity.
   Examples: teach a course, conduct research, train and mentor students, lead a small group or lab, serve on a university or student committee, volunteer for community
2. List the specific tasks associated with each activity.
   Examples: draft syllabus, conduct office hours, grade exams
3. List the skills developed through each task.
   Examples: writing, time management, critical thinking

Next, reread this list of skills and consider which you like using most. That is, which skills would you like to use in your next job? If there are skills you enjoy using that do not appear in your responses, write those down as well.

Exercise: Skills through Reflection

Another way to consider skills that you have built is to think about skills that others have recognized in you, those that bring you energy, or those that make you feel confident. Reflect on the following questions and write your responses in your laptop or career notebook. Take into account not only your academic work but other activities in your life, such as community work, hobbies, interactions with family members and friends, and so on. Write whatever comes to mind instead of trying to narrow your responses.[5]

1. What compliments or other positive feedback have you received for particular activities? Positive feedback may be as simple as a smile or as significant as a pay increase. Write down the compliments, briefly explaining each situation.
2. When have you felt the most alive and energetic? List specific situations. Examples: giving a presentation and getting rave

reviews, running in a 10K, planning and preparing food for a dinner party.

3. When have you felt the most confident and capable? List specific situations. Examples: passing your qualifying exams with honors, having someone ask for your ideas or advice about a particular technique.

Share these stories with a partner or friend, asking them to reflect on skills built in each as you reflect on the same. List skills that emerge from your conversation.

Taking a look at the skills generated in the two exercises above, what common skills emerged? Which skills are most important for you to use in your next job? In identifying which skills belong in this final group, it is important to remember that *skills do not equal interests*. That is, you may be gifted in a particular area but not enjoy using its associated skills at all. Others, seeing your talents, may encourage you toward listing these skills as assets, but you should list only those skills you *enjoy* using.

You can now apply the list of skills you have identified to assess your suitability for a variety of careers. For example, I once worked with a woman who had a PhD in cognitive psychology. I used her responses from this exercise to brainstorm careers away from the bench and the classroom. She had lots of experience teaching and conducting research, but she recognized that she was much more interested in—and good at—detail work, including computer programming. She decided to consider a career in data science. After spending many months developing a network in that field, she was able to obtain an interview for a job she ultimately secured.

Remember: skills do not equal interests . . . but if interest exists, skills can be built.

In Chapter 3, you will be asked to list your skills for a career development profile. Taking these skills into account will help you to ascertain whether you need additional training to enter your careers of choice. Not only is this skills list important for exploring career options, it is also crucial in marketing yourself effectively, both to networking contacts and to potential employers. Whether on your CV or resume or through in-person interviews, being aware of what you have to offer an employer is critical to your success.

CHAPTER 3

How to Identify Your Personal Values

Taking a closer look at yourself before embarking on a job search is the key to success, satisfaction, and long-term stability in a job. Although a job that aligns with your interests and skills is clearly a worthwhile goal, individuals typically thrive in careers that are also a match with their personal values.

In American culture we rarely slow down enough to do the kind of introspective work required to identify our most salient values. Necessity often drives job searches in the United States. If you are graduating in a few short months and your PhD advisor offers you a postdoc, the position may meet both your need to find something quickly and your need for an income. However, if you can carve out time to reflect on a brief exercise about values, chances are you will be able to find a position that may suit you better, and you won't feel ill-equipped to make a decision if a job happens to come your way.

Personal Values and the Workplace

Is a flexible schedule most important to you and your family? A high salary? Frequent travel? Helping others? These are questions that you may have considered in a cursory way in the past, but the more focused and deliberate you can be in identifying your values in advance of a job search, the greater the chances you will feel truly satisfied in the position you ultimately accept.

Exercise: Values Clarification

In this exercise you will focus on your personal values. As you rate each value, think about how important the value is to you in a *work setting*. Keep in mind that there are no right or wrong values; it is a process of identifying what matters most to you rather than to someone else.[1]

Rate each of the following values on a scale from 1 to 5, with 5 being a most important value to you and 1 being not important at all.

_____Advancement	Be able to get ahead rapidly, gaining opportunities for growth and seniority from work well done.
_____Adventure/ Risk-taking	Have duties that involve frequent physical, financial, or social risk-taking.
_____Aesthetics	Be involved in studying or appreciating the beauty of things or ideas.
_____Affiliation	Be recognized as an employee of a particular organization or institution.
_____Altruism/Help Society	Do something directly or indirectly to contribute to the betterment of the world or a greater good.
_____Balance	Have a job that allows time for family, leisure, and work.
_____Challenge	Engage with complex questions and demanding tasks, solve difficult problems.
_____Community Activities	Become active in volunteering, politics, or service projects.

____Competition	Engage in activities that measure my abilities against others.
____Creative Expression	Be able to express my creative ideas in the arts and communication.
____Creativity	Create new ideas, programs, organized structures, or anything else that is unique and novel or not following a format developed by others.
____Competence	Demonstrate a high degree of expertise and mastery of job skills and knowledge.
____Excitement	Experience a high degree of stimulation or frequent novelty and drama on the job.
____Fast Pace/ Time Pressure	Work in circumstances where work is done rapidly and there is little room for error.
____Financial Reward	Earn a larger than average amount of income.
____Flexibility	Work according to my time schedule.
____Friendships	Develop personal relationships with people as a result of work activity.
____Fun	Have opportunities to be playful and humorous at work.
____Harmony/ Tranquility	Avoid pressures and stress in job role and work setting and seek harmonious relationships.
____Help Others	Be involved in helping or being of service to people directly, either individually or in groups.
____Independence/ Autonomy	Be able to determine the nature of work without significant direction from others; not have to follow instructions or conform to regulations.

____Influence People   Be in a position to influence attitudes or opinions of other people.

____Knowledge/   Develop new information and ideas.
Research   Engage in pursuit of knowledge, truth, and understanding.

____Leadership   Direct, manage, or supervise the work done by others.

____Location   Live somewhere conducive to my lifestyle, leisure, learning, and work life.

____Make Decisions   Have the power to decide courses of action or policies, or make decisions regarding the work activities of others.

____Moral/Spiritual   Feel that my work is consistent with my
Fulfillment   ideals or moral code.

____Personal Growth   Have work that enables me to grow as a person.

____Physical Challenge   Have a job that requires bodily strength, speed, dexterity, or agility.

____Public Contact   Have a lot of day-to-day contact with people.

____Recognition   Get positive feedback and public credit for work well done.

____Routine   Have a work routine and job duties that are largely predictable and not likely to change over a long period of time.

____Security   Have a stable work environment and a reasonable financial reward.

____Status/Prestige   Gain the respect of friends, family, and the community by the nature and level of responsibility of my work.

____Teamwork   Have close working relations with a group; work as a team for common goals.

____ Variety	Have a wide range of work responsibilities frequently changing in content, setting, people, and activities.
____Work Alone	Do projects by myself, with little contact with others.

Having rated each value above, go through and circle all of those values you rated as a 4 or 5. Then put a star next to the top nine values most important to you.

Next, list your top nine values in priority order, with 1 being the most important, 2 being the second most important, and so on.

Take a close look at the values you have listed. Can you imagine what it would feel like to be in a job that met all nine of these values? As fulfilling as you might imagine that kind of job, it's more likely that you will have to make some compromises on one or more of these nine values as you progress through your job search. For example, what if you were faced with removing three values from this list? Some PhDs I have worked with in the past have struggled with this decision. However, it is worth taking your time and determining which values on this list are most important to you.

If you were to remove three values from your list of nine, what would your new list of your top six values look like? As before, list these in priority order, with 1 being most important, 2 second most important, and so on.

What if you had to further narrow the list? Try taking away three more values. This step tends to be either very easy or very difficult for most. Determining the top three most important values to you in a job is an exercise PhDs rarely give themselves the chance to do.

List your top three values in priority order, then consider whether it was difficult or easy to come up with your final list of three values. Perhaps this exercise revealed to you the importance of trying to incorporate your top six values into your next job. Perhaps it was easy for you to come up with your top three values, but more difficult to prioritize the list of nine.

Let's reconsider your list of nine values. Can you once again imagine what it would be like to be in a job that met many or most of these values? I like to refer to your list of nine values as your "ideal." If you can find a job that meets all or most of these values, great! While I refer to your top nine values as your "ideal" list, I call your top three values your "must-haves." I use this descriptor because I think it would be very difficult to feel satisfied in a job that did not satisfy at least your top three values. Keep this list with you as you move through the career-planning process, and compare opportunities and institutions to this important list as you encounter them. Using this list during the career exploration stage will make your career research efforts more efficient and will assist you in making a career decision.

Keep in mind that with every job search and offer you might consider, there will be trade-offs. It is crucial to go into the career exploration stage with an open mind and realistic expectations, knowing that all of the values that you listed as priorities may not be met.

Additionally, it will be important for you to keep the lines of communication open with your partner or spouse throughout this process. There may be areas of importance to you, such as financial reward, risk-taking, or balance that immediately affect your partner and family, and as such it may be helpful to have your partner or spouse go through this same exercise and then compare results.

Recognize that your values may change over time, so reflecting on this exercise every so often may serve you as you progress through your career.

Using Values to Determine Fit

After a workshop in which I administered this values exercise, I was contacted by a PhD in biochemistry. He was interested in discussing his results with me individually, so we set up a time to talk. As it turned out, his top three values were balance, location, and teamwork. Spending time with his family was paramount, and being in one particular geographic location was a priority that he and his wife had long since discussed. These findings were crucially important for him to discuss in relation to careers: he had been considering a career in consulting. He was distressed when he came to see me because he understood that the top consulting firms required very long work weeks and significant travel.

Having values that seemed to be in conflict with his career of choice did not necessarily mean that he would have to abandon that field. He did, however, need to conduct more research to find a firm that offered a more flexible schedule in the geographic region he was targeting. His research led him to a few firms in his location of interest, and through connections on LinkedIn, he was able to able to find more information about the organizational culture of those firms and their potential fit for him.

Drafting a Career Development Profile

In addition to values, PhDs must consider several factors to make career decisions, including time left to degree or to the end of a post-doctoral appointment, geographic preference, visa status, family needs, and more. Reflecting on these questions and on work you have done in prior chapters, you can now craft a career development profile that is a fairly accurate reflection of yourself and your priorities. Your profile will allow you to consider career options and to set goals accordingly.

Together, your responses to the exercises you completed in Chapters 1–3 will give you a clearer picture of yourself as a scientist and as a potential contributor to the workforce. To create your career development profile, collate your answers to those exercises in one place.

1. List your top three interests from the exercise in Chapter 1.
2. List the top three skills you enjoy using most, identified in Chapter 2.
3. List your top three most important values, prioritized in the exercise above.

Take a look at the profile you have created. What career fields or occupations come to mind as you read through your profile? For this brainstorming exercise, try giving your career development profile to a friend or colleague. Have them read it and give you two or three potential areas for further exploration. Keep those in your laptop or your career notebook.

Creating your profile will allow you to think more critically about the careers we'll discuss in Chapters 4–6. Refer often to your profile as you explore different careers and as you come closer to making a career decision.

PART TWO

Getting to Know the World of Work

CHAPTER 4

To Postdoc or Not to Postdoc?

As competition for faculty positions has grown increasingly fierce, postdoctoral training has become a requirement for securing a tenure-track position in most scientific disciplines.[1] Postdoctoral training may also be required for research positions in different sectors, such as industrial or government labs. Still, it is important to recognize that postdoctoral training is *not* required for many jobs in science. It is worth considering this next step carefully, before searching for a postdoc appointment, rather than jumping directly into this position without a long-term career goal in mind.[2] One blogger spells out the issue:

> I'm in the final few months of my PhD and am deciding whether or not to pursue a postdoctoral fellowship. I am 30 years old and am married with no kids. I conducted my graduate studies in an excellent lab, have produced several high-quality papers (including one first-author paper in *Science*) and won some awards. I just received an NSERC [Natural Sciences and Engineering Research Council of Canada] Postdoctoral Fellowship, which is 2 years of postdoc funding that can be used at any lab in the world. To an outsider, the world is my academic oyster.
>
> There are many reasons cited to undertake a postdoc (to travel, to see if academia is right for you, because you can't find a "real"

job, to live in the same city as your partner, to kill time, because you got a fellowship). For me, however, the purpose of a postdoc is to gain new skills and experience so that I can get a job as a professor. To that end, I am not interested in pursuing a postdoc simply because I can. I want to make the decision now: either I want to be a professor and therefore need to complete a postdoc or I am leaving academia and therefore need to start looking for jobs. I am either going to go for it or get out now.[3]

This point is an important one: *a postdoc is not a career goal.* It is instead a training period during which you can develop the skills, knowledge, and experience needed for your ultimate career goal. The postdoc should not be viewed as the terminal step in your career path but as a stepping stone, useful only if it is required.

Is a Postdoc Required?

According to the research I conducted through my survey project, "Identifying Career Pathways for PhDs in Science," 85 percent of all currently employed PhDs in the sample (N = 3,838) stated that a PhD was required or preferred for entry into their current position. However, when asked whether postdoctoral training was required or preferred to enter their current job, only 40 percent responded that a postdoc was either required or preferred.

In addition to understanding whether a postdoc is required for your intended career field, you need to think carefully about how you would like to spend the next several years of your life, and about any financial commitments you may have. Do you have student loans? Are you providing financial support for your family? Are you willing to live on postdoc wages for the next three to five years, rather than moving into a full-time, permanent, most likely higher-paying job? You need to consider the financial implications of entering a postdoc, as well as the career implications.

In her book *How Economics Shapes Science,* the economist Paula Stephan breaks down the loss of potential earnings of PhDs in the biological sciences versus those who earn a master's degree in business administration. She estimates a loss of $1,219,257 in lifetime earnings for those who spent seven years completing their PhD in biological sciences compared to the holders of MBAs, and an additional loss of $109,124 for those PhDs who spend an extra year completing their degree. For those who complete their PhD in seven years but also complete a three-year postdoc, the loss in lifetime earnings is even greater: $1,272,680. She notes that the difference in earnings between the MBAs and the PhDs makes "it quite clear that reasons other than money enter into the decision to pursue a career in science and engineering. If it were only money, virtually no one would choose such a career."[4]

Stephan focuses on wages lost through increased time to earn a degree and an increase in the tendency to take on a postdoc. Naturally, interests, skills acquired, and personal values play into any individual career choice. Still, Stephan's point is well taken in demonstrating the choices—and potential losses—that PhDs face when trying to discern a career direction, and in particular, whether or not to engage in postdoctoral training.

Another example I like to share with PhDs considering whether or not to apply for a postdoc illustrates the mean salaries by educational level for people ages 25–44 compared to a starting postdoc salary (Figure 4.1).

To be sure, a postdoc is still required for positions in tenure-track faculty research jobs and for most college teaching positions, and many postdocs engaged in training hope and expect to progress toward these positions. According to the data set I amassed in 2015 in my survey project, 81 percent of current postdocs (N = 3,247) named "university faculty" as their long-term career goal, with a focus either on teaching or on research.

To enhance their chances of being among the small percentage of candidates who are selected for tenure-track positions, some PhDs take on a postdoc to increase their number of first-author papers. Some seek to apply for transitional grants, while others apply for training positions to learn new techniques, to gain teaching

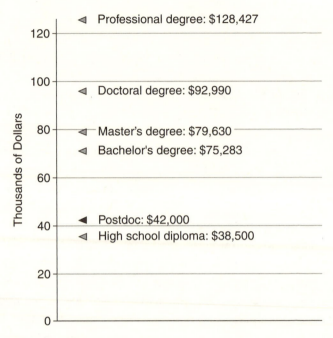

4.1. Mean salaries by educational attainment and age
US Census, Current Population Survey, 2014, Ages 25–44.

experience, or even to change their research focus completely. These may use a second postdoc to retrain in an area that will make them more competitive for jobs.

Most of these decisions were predicated on the knowledge that the academic job market is now more competitive than ever. Generally speaking, if you are contemplating a long-term career as a faculty member at the college level, postdoctoral training is essential to be competitive. According to data acquired through my research project on career options for PhDs, postdoctoral training was required or preferred for 70 percent of those respondents currently in a tenure-track position or tenured (N=901). Although this is a relatively small sample size, it does demonstrate that a small number of candidates move directly into tenure-track positions without additional training.

Outside of the tenure track, it is much less likely that a particular position will require postdoctoral training, though there certainly are instances where PhDs can use such experience to enhance their candidacy. If you intend to work in a research setting for a long period of time, a postdoc may help you gain greater autonomy and learn how to manage groups. It may increase your number of peer-reviewed publications. Employers that might find postdoctoral training attractive include government agencies, nonprofit research organizations, research-focused corporations, healthcare centers, and other organizations driven by scientific research.

Here is a sample job description for a staff scientist. Some job descriptions are not as detailed and indicate simply a certain number of years of work required beyond the PhD, such as this job ad.

Summary:

This position will be responsible for leading and performing complex in vivo and ex vivo tumor studies in mice using biospecific antibodies as part of the immune-oncology efforts within the Oncology and Angiogenesis group. The qualified individual will have a strong background in one or more areas of immunology, oncology and/or immune-oncology.

Technical skills in tumor cell biology, in vivo tumor growth and model development, ex vivo immune cell analysis (FACS) and cell-based assays are essential. Experience with Immuno-PET or radioactive imaging in mice is a plus. Attention to detail coupled with an aptitude for learning are also key elements of the position. Excellent communication and management skills are also required as this candidate will be expected to work collaboratively and communicate well both within group and across group settings.

In addition, the candidate must be able to effectively collaborate with colleagues in both the oncology and antibody development groups, writing reports and communicating results in a timely manner to other members of multidisciplinary teams. The successful candidate is expected to work creatively and independently in a dynamic team environment. The ideal candidate should be well organized with excellent interpersonal and oral/written communication skills and able to manage several projects simultaneously. Other requirements: strong analytical and critical thinking skills and attention to detail.

Responsibilities:
- In vivo tumor growth and monitoring by caliper, IVIS, and other means
- Dosing agents to mice by s.c., i.p. and i.v. routes
- Ex vivo tumor and peripheral LN, spleen and blood analysis by FACS, MSD
- Complex in vitro cell assays (cytotox, immune stimulation)
- Purification and analysis of immune cell subsets from mouse organs and blood and human blood
- Advanced written and oral presentation skills
- Highly organized with attention to detail
- Potentially manage 1–2 research associates
- Ability to write documents to support IND filing, patent applications and external conference posters and papers

Experience and Required Skills:
- PhD with 3–5 years additional years of experience in research
- In vivo tumor growth (immunocompetent and immunocompromised mice) and monitoring by caliper, IVIS, and other means
- Dosing mice with agents s.c., i.v. and i.p.
- Mouse handling and dissection of peripheral LN, spleen and BM
- Purification and analysis of immune cell subsets and cytokines from mouse organs and blood ex vivo using FACS, IHC and MSD/Luminex
- Advanced written and oral presentation skills to varied audiences

- Highly organized with attention to detail
- Management experience a plus
- Ability to write documents to support IND filings, patent applications and external conference posters and papers

Generally speaking, if a research position requires years of experience in research training beyond the PhD, postdoctoral training may suffice.

Fields that Do Not Require a Postdoc

If you are interested in jobs within university administration, business, consulting, writing, data science, public health, museums, K–12 teaching, or myriad other fields, you most likely will not need postdoctoral training to gain entry if you have spent the time and energy necessary to gain requisite skills and experiences.

To determine whether a postdoc is in fact required for your field of interest, contact someone who works in that field and ask this question directly. You will also find data on degree level and training required for different fields in Chapter 5. It is essential that you learn more about various fields now, *before* entering a postdoc. I have worked with hundreds of postdocs who found that their training after completing their PhD was not required for the field they ultimately chose.

If you are currently employed as a postdoc and interested in a field that does not require additional training, do not despair! Chances are you have been amassing additional skills and training through your postdoc work that are valuable to almost all employers, including the ability to work independently. Still, the more you speak with PhDs in different professions, the more you will learn about what types of experiences are most meaningful and would make you most attractive for positions in that area.[5]

Types of Postdoctoral Appointments

If you determine that in the long run it would benefit you and your job search to engage in postdoctoral training, you may wish to learn about different types of postdocs, as well as the benefits and drawbacks of engaging in varied training settings.

Academic Postdocs

Most postdocs in the United States are employed in the academy—in 2013, at least 75 percent of them.[6] Their popularity, perhaps, can be attributed to several factors: academic postdocs are easiest to find; they are the most traditional type, dating back to the apprenticeship model in German universities; postdocs in other settings may be less visible to graduate students; and academic postdocs are assumed to leave every future career door open.

There are surely many other reasons for their attraction, and those may include opportunities to publish, to write grants, to teach, to exert a significant degree of autonomy, and so on. The downsides of academic research as a postdoc can include isolation, low salary, and mediocre benefits.

Still, many PhDs may see academic postdocs as a natural extension of graduate school. With academic postdocs being easy to secure, they may appear to be the most immediately accessible option.

Industry Postdocs

The number of industry postdocs, at least according to the National Science Foundation, is still relatively small, with just 14 percent of scientists and engineers across all disciplines engaged in postdoctoral training in an industrial setting in 2013.[7] The culture of each particular industrial environment may vary, but generally speaking the focus on teamwork is pervasive, and the work largely mission driven. Some companies habitually hire former postdocs as permanent employees following a postdoc in industry, while others avoid continuing the relationship.[8] Whether hired by the same employer or not, exposure to a corporate environment can be seen as a plus by most industrial employers interested in hiring PhDs.

Another perk of working in an industry postdoc is access to cutting-edge technology and state-of-the-art facilities. You may find the same at an academic, government, or nonprofit research institution, but it is more likely to be the standard here. Along those same lines, salaries tend to be higher—sometimes significantly so—for industrial postdocs. One drawback to working in an industrial setting, though, is that it may be more difficult to publish your work. While this is not always the case, it is true in some corporate settings. Also, you may not have the ability to write grants, if that is an area of interest for you.

Q: Should I choose an academic or industrial postdoc if I think I might want to return to academic research someday?

A: It depends on the specific postdoc, program, project, and environment. Some industrial postdocs are highly academic, collaborating with local academic institutions and encouraging publication, whereas others closely guard intellectual property and do not publish results as readily.

Government Postdocs

Postdoctoral training in government agencies is less often considered, perhaps because participants make up only 10 percent of the postdoc population in the United States, according to National Science Foundation data.[9] Nonetheless, some argue that postdoc work in a government agency may incorporate the best parts of academic and industrial postdoc work: publication is typically encouraged; salaries are usually higher than those of an academic postdoc; and materials and equipment can be readily available. Another perk for foreign scholars is that, unlike some corporations, many government agencies, including the National Institutes of Health, hire international scholars. One drawback of this environment is that

SPOTLIGHT ON INDUSTRY POSTDOCS

Postdoctoral training programs in industry vary greatly, depending on the organization. One company that hosts a structured postdoctoral program is the Novartis Institutes for BioMedical Research (NIBR), the research arm of Novartis International AG.

A key feature of the postdoc program at NIBR is that together with their mentors, postdocs identify and work on a fundamental research question and aim to publish their results in leading journals. This is not necessarily the case for all industry postdoc programs and is a point of clarification for those interested in these types of positions, as having publications may help enable your search for employment following the postdoc. For the NIBR program, publications are an important aspect of supporting flexibility in their postdocs' future research careers in academia or industry.

Of the approximately 350 postdoc alumni to date from NIBR, 70 percent have gone on to positions in research: 58 percent have taken full-time positions in pharmaceuticals or biotechnology (mostly at other companies), 4 percent went on to faculty positions in academia, and 8 percent to another postdoc position. The remainder moved on to nonresearch, science-related, or other positions. Asking about career outcomes is critical for those exploring industry postdoc programs, as some companies may tend to hire their own postdocs for permanent positions.

NIBR has a strong desire to foster collaborations with academic scientists, and the postdoctoral program is essential to these efforts. Each postdoc is encouraged to engage an academic advisor, in addition to working with their primary mentor at NIBR. Additionally, NIBR is launching a highly competitive Young Investigator Award that provides start-up funding for their postdocs who go on to faculty positions.

With research being conducted in cancer biology, neurobiology, pathways biology, and regenerative medicine, among other areas, NIBR offers postdocs the opportunity to broaden their experience, build scientific and collaboration skills, and become more agile thinkers. Additional benefits include access to investigators across multiple sites, a wide range of resources and expertise, and a portfolio of courses and workshops. At the same time, postdocs bring fresh perspectives to advancing innovative research at NIBR.

government postdocs typically do not have the ability to write grants, which is an absolute prerequisite for faculty positions, and may or may not have the ability to teach, another requirement for faculty positions, at least in teaching-focused colleges.

Teaching Postdocs

If you have not had much (or any) teaching experience in your graduate program and would like to develop some classroom experience, investigate teaching postdocs across the country. These fellowships typically combine scientific research and classroom teaching at a local college. These fellowships also tend to include workshops and training in pedagogy, classroom management, and other resources needed to be an effective teacher. A list of teaching fellowships is compiled by the National Postdoctoral Association (NPA).[10]

Field-Specific Postdocs

If you are interested in securing positions in a few select additional fields, such as science policy or technology transfer, entering a specialized postdoc position to build skills in these areas will certainly make you more attractive to potential employers. These training positions are not required to enter these fields, but could provide concrete experience and skill-building in writing, oral communication, and perhaps most significantly, networking opportunities. (You may also gain the skills and experience necessary for entry in these careers during your PhD, which we will discuss in Chapter 6.)

There are some specific fellowship programs available, with one of the most widely known being the American Association for the Advancement of Science (AAAS) Fellowships. These fellowships, like teaching fellowships, engage postdocs in training for particular professions. Selected fellows may work in science policy with congressional officials or in federal agencies, or may work in mass media as reporters, editors, or production assistants.[11] These fellows engage in the field, meet professionals, and contribute their research skills, critical thinking skills, analytical skills, and communication skills to serve a variety of organizations. There are also opportunities available in specific government agencies, such as the National Cancer Institute, whose Technology Transfer Center offers two different types of fellowships for candidates interested in technology transfer.

Self-Made Postdocs

Given immediate access to information and continuously growing social networks, many students and current postdocs have fashioned their own postdoctoral fellowships by reaching out to organizations directly. Some have found opportunities to work for pay in start-up companies. Some have volunteered to work without pay for boutique consulting firms. These postdocs possessed an entrepreneurial spirit and found success by responding to an internal drive to enter a career, knowing that they had to build their own experience from the ground up. For more information on creating a transitional experience for yourself, see Chapter 6.

Transitional Postdocs

Some PhDs will remain in their doctoral research groups, departments, and institutions as postdocs after completing their degrees to gain more time to search for a permanent job. Although this may be a necessary step for you in your career development, or for your family or partner, it is wise to avoid studying with the same mentor for your postdoc and your PhD. Many employers, including academic search chairs, department heads, managers in industry, and more want to see that someone searching for a research position has stretched their knowledge and taken risks or found new projects by moving into an entirely new research group. If you must stay for some period of time, then you must, but if you know that your time for whatever reason will extend beyond a few months, find another postdoc, even at the same institution, to take on new challenges and demonstrate your teamwork skills.

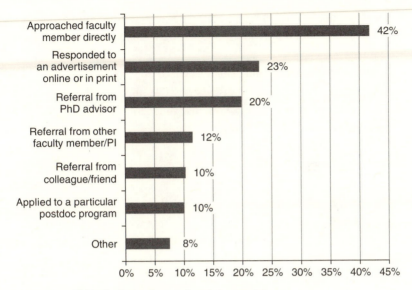

4.2. How PhDs find postdoctoral training positions

Searching for a Postdoc

There are several strategies you can use to locate and apply for post-doctoral positions.

According to data collected through my 2015 study, Figure 4.2 indicates the frequency with which postdocs used the methods listed to find their positions.[12]

Apply Directly to a Faculty Member or Principal Investigator

The most common method for applying for a postdoc is to contact a faculty member or principal investigator (PI) conducting research in a field that appeals to you. This contact most often happens via email, as in this sample:

Dear Dr. ___(last name)___,
I am currently a PhD student enrolled at ___(institution name)___ and will finish my degree in ___(discipline)___ on ___(date)___. I am interested in

gaining more experience in ___(PI's area of focus)___ and am writing to inquire about a potential postdoctoral training position with you.

My CV is attached, and I can be reached at ___(your cell number)___ or via this email address to discuss possible opportunities to join your research group. Thank you in advance for your consideration.

All the best,

___(Your Name)___

This kind of direct contact happens all the time, so do not hesitate to reach out directly to faculty members whose research excites you. If they do not have funding, they will let you know, but those faculty who do have funding available and find your CV compelling will most likely reach out to you.

Newer faculty may even post available postdoc positions on their websites, which takes some—but not all—guesswork out of wondering whether funding is available. (Note that not all faculty keep their websites up to date, so it is still worth sending a brief email.) Additionally, departments may list ongoing postdocs on their websites, but funding available from a particular faculty member may vary.

If faculty members are interested in interviewing you, they will contact you directly. Some of these interviews take place by phone or Skype (or similar), others in person. I discuss preparation for these and other interviews later in this chapter and in Chapter 10.

Apply to Postings in Journals

Another source for postdoc positions is scientific research journals. Major journals and news outlets such as *Science, Nature,* or the *Chronicle of Higher Education,* for example, list available postdoc positions. You should also check the websites of journals in your particular discipline. However, none of these sources will have a comprehensive listing of all postdocs available, so it will still serve you to reach out to people directly if you do not find any listings in your area of interest.

Apply to Specific Postdoctoral Training Programs

Postdoctoral fellowships designed to assist you in building skills in a specific area may be attractive to you as well, depending on your career goals and developmental needs. (See the discussion earlier in this chapter on specialized postdoctoral fellowships.)

Additional postdoctoral fellowships are available for researchers from underrepresented groups. Minoritypostdoc.org is the most comprehensive site, listing fellowships for women; minority groups historically underrepresented in science; the lesbian, gay, bisexual, transgender, and queer community; and others interested in pursuing postdoctoral training in a scientific discipline.

Interviewing for and Selecting a Postdoc

Once you have interacted with a potential PI, you may be invited to interview for a postdoc position. Whether the interview takes place on the phone, in person, or over the web, it is critical that you prepare and engage in deliberate reflection about what is most important to you prior to the interview date. Knowing yourself, your work style, and what you need to be successful in a training role are a few of the areas you should explore before meeting with a PI. Think carefully about both the content of the project (your research goals) and your own professional development (your career goals) and be ready to outline these for the prospective PI. You must have a clear sense of the following *before* interviewing for a postdoc:

1. What do you hope to accomplish by entering this postdoc? That is, what are your long-range career goals, and how does this postdoc fit into those goals?
2. What technical or research skills do you need to build, and will you be able to acquire them through this training?

3. Do you have a definite research project in mind that you want to undertake?

In preparation for your postdoc interview, you should also read through the following questions and have them with you at the interview. If these topics do not arise organically during your conversations with PIs or group members, be certain to ask them yourself.

Questions for Your Faculty Advisor or Principal Investigator

1. How long is the postdoctoral training appointment? If it can be renewed annually, on what does this renewal depend?
2. Does the PI have adequate funding to support you throughout the proposed research project?
3. What are the PI's expectations for the first month? Six months? First year?
4. What is the PI's policy on authorship? On ownership of ideas?
5. What is the size and composition of the research group?
6. How often does the group meet? Are presentations required of each member, and if so, how often?
7. Will you be expected to train and mentor others, or take on a summer undergraduate student?
8. How often can you expect to meet individually with your PI?
9. Does the PI support collaborations with other groups, departments, and institutions?
10. Will you have the opportunity to—or be expected to—write a fellowship application?
11. Will you be able to travel to meetings to present your work? If so, is the number of meetings limited?
12. Will you have the opportunity to teach, if you are interested in doing so?
13. Will you be able to attend seminars and workshops on career and professional development topics? Are there any restrictions?

14. Will you be able to take any part of your research project with you after your postdoc ends?
15. What services exist at the institution to support your career development and job search?

By gathering answers to these questions, you will be better able to make an informed decision about which postdoc position is the best fit for you. I cannot emphasize enough how critical it is to collect this information *before* you begin a postdoctoral appointment, rather than after. In addition to collecting your own information, be sure to speak to current and past group members about their experience. Knowing more about the postdoctoral position you are about to enter, as well as the work style and habits of your new PI, could potentially save you time, money, energy, and stress.

You will also want to investigate benefits available to postdocs at your potential research institute. This information is available from your department administrator, the institutional postdoc office, or the human resources department.

Making the Most of Your Postdoc

Once you have carefully weighed your options and have chosen a position, request the specifics of the appointment *in writing*. This is a step that many graduate students omit, simply because of their lack of experience in the work world. It is totally appropriate to ask your future PI for a letter of appointment that spells out what you have agreed upon. This letter should include:

- your start date;
- your end date;
- your funding source;
- salary and benefits;
- the project you will be working on; and
- any other salient information you think should be captured on paper.

You may need to sign a copy of your appointment or offer letter, or simply reply by email that you accept the position. After that, take the following steps to ensure that your postdoc appointment will be as productive and enjoyable as possible.

Contact Your Institution's Postdoc Office

In 2014, there were 167 offices across the United States intended to support and serve the needs of postdocs. According to a study conducted by the NPA in 2014, 84 percent of these offices offered career development programs and workshops, 77 percent offered networking events, and 59 percent offered individual career counseling appointments.[13] Some offices also offer information about your appointment, including benefits and institutional policies around stipend levels, and will connect you with other resources across your institution. To make the most of your appointment, it will be important to visit this office, get to know the staff, and take advantage of all of the resources available to you. It will also serve you to share feedback with this office about emerging needs of the postdocs at your institution, for these offices are created to serve you.

Before arriving, find out whether your institution has a postdoc office and reach out to them for information on getting started. These units typically have websites designed to assist scholars with adjusting to the institution and the surrounding community. If you are an international scholar and have accepted a postdoctoral position, begin by contacting the international office at the institution. This group will be able to assist you with visa questions and applications, and can help you prepare to enter the country.

The postdoc office may also assist you in finding housing, setting up your benefits, creating an email account, and any other tasks that can be completed before you arrive. If your institution does not yet have staff specifically devoted to assisting postdocs, you might try reaching out to individual units, such as the housing office for relocation assistance, the human resources department or benefits office, and the information technology center.

Map Out Your Research and Career Goals

Once you are settled at your institution, take some time to map out your research and career goals. What is most important for you to accomplish in the first month? In the first six months? First year? You should sketch out a plan for yourself for these time periods and be ready to make some adjustments once you speak to your advisor.

When planning your work time, think first about your long-range goals. What would you like to have accomplished by the end of the year? If you want to have completed your first round of data collection, work backward and set monthly goals. For example, you might assign yourself the task of reviewing relevant literature for the first two months of your appointment. For the third, you may start to sketch out your research design. Again, your plan needs to be flexible, based on the expectations of your PI and the activities of the research group.

Once you have mapped out monthly goals, move to setting weekly short-term goals that are specific and measurable. For example, you might decide that you need to read and take notes on three articles per week by Friday at 5:00 PM. This goal is measurable, time-delimited, and specific. If you have not completed this task by the appointed time, you will have to move it to the following week.[14]

Go through this same exercise with your career goals. Where would you like to be in ten years? In twenty? If you are unsure, you can still craft a long-term goal related to your decision. For example, you may decide that you need to make a decision on a career path by the end of your first year. Again, work backward from there: setting concrete, short-term goals every month will help you gain a clearer view of your long-term goals. From there, setting daily goals will help you make steady progress. Once you have committed these plans to paper, share them with your PI and schedule a one-on-one meeting to discuss them.

The document outlined above can be referred to as your Individual Development Plan, or IDP. Although individual development plans have been in use in organizations for several decades, they were introduced to the scientific community in a formal way by the Training and Careers Committee of the Federation of American Societies of Experimental Biology (FASEB) in the fall of 2001.[15]

Presently, several federal agencies that are substantial funders of postdoctoral fellowships are encouraging research institutions to submit their own IDP procedures as a part of the grant review process. Clearly, setting goals for both your research and your career are paramount not only to your success, but also to that of your particular institute, and to the scientific enterprise in general. For more information on constructing your own Individual Development Plan, see Chapter 8.

Discuss Your Plan with Your Advisor

Once you have drafted an IDP, schedule a meeting with your advisor to discuss it. Conflicts that arise between PIs and their postdoctoral trainees tend to be based on miscommunication and unspoken expectations—from both sides. It is crucial for your development, both as a research scientist and as an emerging professional, to communicate effectively with your advisor (indeed, with everyone you intend to conduct research with), and to understand clearly the expectations for your work.

When meeting with your advisor to discuss your draft IDP, be open to new directions, new possibilities, new ways of thinking and working. Your PI may have an approach to a particular problem that may require skills you have not developed, for example. This is just one reason why it will serve you to clarify expectations, and to communicate early on your questions or concerns about your research project. If you are unfamiliar with a technique, for example, will your PI connect you to another group member who has those skills? Will your PI allow sufficient time for you to develop those skills before you can really get your project up and running? If both of you agree on short-term and long-term goals for your work, you will be more likely to enjoy a healthy working relationship and more likely to succeed in your work.[16]

Discussing your career goals with your advisor may be more difficult than talking about your research. What's more, you may not be entirely sure of your long-range career goals when you first meet. Still, you will benefit from sharing your career thoughts with your PI sooner rather than later to avoid misunderstandings. You will also be able to ask about participating in activities you know will serve you. Would you like to enroll in a grant-writing course? Would you like to teach? Create a blog for your research group? Present at an upcoming meeting? Explore industrial collaborations? Do not be afraid to initiate this conversation if you are committed to pursuing a particular career: you will need time to develop the relevant skills and experience. If you cannot speak openly with your advisor about your career goals, it may be worth exploring other opportunities. You have to recognize your needs and take charge of your career.[17]

Build Your Skills and Experience

Use the postdoctoral training period to advance yourself professionally. Guided by your IDP, identify opportunities to build skills and experience. If you want to strengthen your leadership skills, volunteer to serve the postdoctoral association at your institution. If mentorship is important to you, request a summer student. Attend seminars in various departments to explore innovative, interdisciplinary approaches to your project. Approaching your work from a different perspective may spark creativity, or generate new partnerships or collaborations. Write a fellowship application. Grant writing is valued in many professions, required for some, and will help you to understand your project more fully and to identify your focus more clearly. This is *your* time to grow as an independent researcher and professional. Use it wisely!

If you are unsure which skills will make you the most attractive job candidate, consider the following: in 2005, the NPA assembled a committee to develop a list of core competencies that all postdocs should possess by the end of their appointment. Their report, published in 2009, includes:

1. Discipline-specific conceptual knowledge
2. Research skill development
3. Communication skills
4. Professionalism
5. Leadership and management skills
6. Responsible conduct of research

Each competency listed is linked to a description and resources on the NPA site.[18]

In addition to skill-building, consider engaging in an experience that will enhance your marketability for a particular job. If you are undecided about your long-term career goal, you may need to conduct more research before jumping into an activity in order to develop a clearer idea of careers you might enjoy. To learn more about different fields as well as transitional experiences that will make you more marketable, take a look at Chapters 5 and 6.

Find Career and Professional Development Programs

Your institution's postdoc office or career center may offer professional development programs. Although career centers were initially conceived to serve undergraduates, almost all now serve current graduate students, and some even serve postdocs. Your institution's human resources office may also offer programs on job searching, interviewing, and polishing your CV.

In addition to formal administrative units at your institution, look for student- or postdoc-led groups or events. Many graduate student and postdoc clubs and associations host valuable career development programs throughout the year. At some universities, for example, graduate students create graduate consulting clubs not only to offer practice case study sessions, but also to connect current graduate students and postdocs with employees from consulting firms. If you do not find any such programs anywhere at your institution, you might consider building one. Not only will you develop organizational, presentation, and leadership skills by assembling a career panel, for example, you will also make networking connections that may be helpful to you.

Become an Active Member of the Postdoctoral and Scientific Communities

Join the NPA, which is free for you if your institution subscribes as an institutional member. The NPA is a nonprofit organization that offers information for and about postdocs in the United States. Their site includes links to career information, as well as sample mentorship plans, news about policy changes impacting postdocs, and data on postdocs and institutional policies. The NPA also hosts an annual meeting with workshops, keynote presentations, and networking opportunities for postdocs and postdoc office administrators.

Many institutions also have active, local postdoctoral associations. These organizations are volunteer groups that typically develop to advocate for postdocs and serve their needs. Many such organizations host career and professional development events throughout the year either live or online. You might consider serving such an organization by hosting an event in an area of interest to you. Might you like to start a women's group? Or perhaps a journal club in your discipline? At some research institutes, a group of postdocs from several different disciplines might decide to start a postdoc science club that meets once a month to share slides on their current research, and then engage in an active, scientific discussion. These kinds of programs are typically well attended and meet with positive evaluations by attendees and presenters alike. Such interdisciplinary clubs benefit participants not only by building leadership and teamwork skills, but also by communicating science in a coherent way to people outside of their discipline.

Other organizations critical to your development as a scientist and a professional include professional scientific societies. Some of these organizations have local chapters with career events and volunteer opportunities. Getting involved with a professional society is one of the most helpful actions you can take to further your career. Not only do these organizations provide wonderful avenues for networking, such as local and national or international conferences, they offer worthwhile leadership activities, such as committee participation, task force work, and so on. For example, some scientists I have worked with have gained valuable organizational and teamwork skills by planning career events through the Association for Women in Science (AWIS). Others have been involved in campus chapters of the Society for the Advancement of Chicanos and Native Americans in Science (SACNAS), the American Chemical Society (ACS), the Society for Neuroscience (SFN), and more. Opportunities for you to develop as a professional abound, even some that can be experienced from your lab bench. For instance, I have participated in many online webinars where I can interact with colleagues from across the country. These interactions have led to service on task forces, and I value the relationships I have built through these activities.

Find Multiple Mentors

> [T]he postdoc has the *right* to expect good mentoring: oversight, feedback, sympathetic consultation, and periodic evaluations.
>
> —National Academies of Science[19]

Throughout my career, I have heard some postdocs express dismay that their "advisors" (PIs) are not "mentors." This is not an uncommon sentiment, so be sure to think through what you expect from your PI *before* accepting a postdoc position. If you are aware of your personal style and what you need to be successful, you will be in a better position to assess whether a potential advisor may also serve as a mentor.

That said, some faculty members are better mentors than others, and the person you choose to study and train with may or may not be the best fit for you as a mentor. Whether or not this person is the perfect mentor for you, you will benefit by selecting multiple mentors who can help you in different ways. First, find another scientific mentor, someone senior to you who understands your science and can communicate with you about it. Try also to find a mentor who is in a field of interest. You may take this person out for coffee (and pay, of course!) to find out more about his or her path. You may then want to find a mentor who can serve as a writing coach, keeping you on track and perhaps reading some drafts of your work. If you are a member of a group that is underrepresented in science, you might consider finding a mentor in the same group, scientist or not, in your discipline or outside of it.

Building relationships with potential mentors may happen organically or formally. You can approach those whose work interests you and ask whether they might be willing to serve in that role for you, or you may simply endeavor to spend more time with them, observing their areas of strength that you would like to emulate or requesting review of your work to gather feedback. Be cognizant of their time, understanding that all mentors have myriad responsibilities in their roles as professionals. Reciprocate in any way you can, offering to assist mentors in their work whenever you see fit. This will ensure a robust, collaborative relationship, rather than one that serves only your needs.

Take Time for Yourself

In addition to tending to your scientific work and progress, it is critical to care for *yourself* throughout your postdoc. Surround yourself with people who support you and your work. Build in time for yourself throughout the week. Burnout in research science is a genuine risk, and it is up to you to carve out time for your interests outside of work—and to commit to this time. Do you enjoy hiking? Knitting? Playing fantasy football? Cooking? Some postdocs I have worked with have found it easier to join a group or class in order to dedicate time to a particular activity outside of work. You might even found a group around a specific interest. For example, a noontime walking club could benefit the entire research community, with faculty, staff, students, and postdocs able to join anytime. You could start something similar at your institution.

Caring for your health is another important piece in ensuring that you can do your best work. If your research is keeping you in the lab fourteen to sixteen hours per day and you find you are not eating well or not getting enough sleep, you might think about talking to your advisor and brainstorming another approach to the project you are working on. Can a piece of it be taken over by a graduate student? Another postdoc? Can you hand over a side project that your PI gave you in the beginning, maybe before you started a larger project? Try to develop some solutions on your own before approaching your boss to lessen the burden of moving tasks around.[20]

Eat well, sleep well, and enjoy your time as a postdoc. Faculty and senior researchers in many sectors have said it was the best time of their lives. Embrace the work, find training in areas of interest, visit the postdoc office, and be sure balance is present in your life. This may assist you in enjoying a successful, productive, and gratifying postdoc experience.

Plan with Your Family and Future in Mind

If you would like to have children, think carefully about the timing. Many postdocs have children during their training, some arrive with children in tow, and still others will wait to have children until after departing from their postdoc and entering a permanent position. So many factors impact this decision, and even with careful thought and discussion, you cannot know the outcome of your planning.

Still, there are some steps you can take to be prepared should you or your partner become pregnant or decide to adopt during your postdoc. First, be sure to carefully examine the maternity and parental leave benefits at your institution. Some postdocs have been unpleasantly surprised to find that there are no accommodations for paid maternity leave or parental leave at their institutions only after they have become pregnant. Your institutional postdoc office, human resources, or the benefits office should have this information.

Next, think carefully about your work. Would you like to continue with your research after taking some period of time off? If so, how much time would you like to take? If you are not continuing your research, do you have the financial reserves to stop working after your child arrives? Regardless of your decision, be sure to share it with your PI as soon as you can. It is your job as a professional to keep your advisor informed of your plans as you make them, recognizing that they might change once the baby arrives.

Having made a decision about work, be sure to think carefully about child care. Caring for an infant is extremely expensive in most regions of the United States and requires careful planning, not only on the financial side, but also on the logistical side. Many child care centers have waiting lists months in advance of when babies actually arrive. Look into care options as a part of your early planning so you will have a better sense of what to expect.

And naturally, it is critical to care for your health if you do become pregnant. I have met women postdocs who have had to ask to be reassigned to new projects once they became pregnant, since the materials they worked with were too dangerous for their growing babies.

Meeting Possible Challenges during Your Postdoc

In spite of adhering to the suggestions outlined here for a successful postdoc, healthy working relationships, and work-life balance, you may encounter extenuating circumstances that impact your training experience. We will explore some of these issues, as well as strategies for coping with them.

Lack of Support from Your PI

This situation may take many forms. In most of the cases I have encountered, this phrase means either a lack of interaction with a faculty advisor, or an advisor's spoken (or unspoken) disappoint-ment in or lack of support for one's career goals. Let's take these two conditions one at a time.

Some faculty travel frequently for their work. In this instance, maintaining a close working relationship with your research advisor can be a challenge. If a PI is unavailable to support your work, find another source of support. More and more, postdocs turn to their peers for mentorship, for sharing of ideas, and for tackling problems that arise with their projects. Additionally, find another faculty member who can serve as a scientific mentor, as described earlier in this chapter.

If your faculty advisor is unsupportive of your career goals, it may be simply a matter of concern on the part of your PI that your work will suffer because you are focused on your career development plans or activities. It is important for you to be aware of your PI's research goals and timeline for your projects, and likewise to com-municate your own research and career goals. Be knowledgeable about and sensitive to your PI's work, tenure plans, grant status, or other issues. Ask how you can help—but be sure to use your time wisely and remain committed to your own career develop-ment. Your postdoc is a job and a priority, but the same is true of your future.

Q: If I don't have a great relationship with my PI or my project is not going well, how long should I stick it out and remain in my postdoc?

A: Pay attention to your intuition. If you are unhappy going to work every day, it may be best to look for a new appointment. Postdocs move around frequently, so do not hesitate to look for another postdoc appointment if you are in a difficult situation in terms of your research, career, or mental health. Employers will likely look more favorably on a six-month appointment followed by a productive two-year appointment than on a two-year appointment during which you have not progressed in your work.

Unexpected Elimination of Funding

You should know when your postdoc funding will expire and plan to spend between six and twelve months searching for a job, depending on your career goals. Sometimes funding ends with the departure of a PI, or expires unexpectedly. In these cases, you need to do a few things to protect your finances. First, ask your PI or department chair whether you can have a few months of transitional time to find a new postdoc. At the same time, take a look at your personal finances. How long can you last on the job market? It is important to build up savings over time for any crises, but if you find you do not have enough to live on, it may be best for you to find another postdoc as quickly as you can. Look locally, and you may be able to find another project similar to the one you were working on.

If you are an international postdoc, be sure to visit the international scholars' office at your institution *immediately,* as soon as you learn of your funding being terminated. This office will assist you in understanding how to deal with any visa issues present.

Termination without Cause

Many institutions have policies that protect postdocs from being terminated without cause and due process. To find out whether this is true for your institution, contact the postdoc office first. If this unit does not exist or is unable to help, contact your human resources office. Finally, be sure to visit your institution's ombuds office. This is a confidential, safe place for you to share the details of your situation and to identify resources available to you.

Loss of a Parent or Loved One

Over the years, I have worked with a few postdocs who have experienced the loss of a parent or other family member during their training period. If you lose a loved one, it is essential for you to take some time to care for yourself. Find the counseling center at your institution, or seek a counselor or therapist outside of your institution. Give yourself time to grieve, but be sure to communicate—or have someone communicate—your plans to your PI. Take the time you need and rest assured that you can recover from such a loss and continue to do productive work.

Has Your Postdoc Been Successful?

A postdoc is never an end in itself; it is always a means to the next step in your career. The greater clarity you have about what lies on the other side of your postdoc (for example, an industry job or a faculty position at a small liberal arts college), the more likely you will be to find one that promotes or supports that goal and prepares you to be the strongest possible job candidate.

A successful postdoc is one in which you use knowledge about yourself and the kind of work you want to pursue, as well as the working conditions and management style that make you most productive, to facilitate agency over your future. You are the chief architect of your future career. Let this perspective guide you throughout your postdoctoral experience, and you are likely to feel fulfilled in your work.

CHAPTER 5

Career Options for PhDs
in Science

What kinds of jobs are out there for PhDs in science?

An examination of current data available in the United States reveals some trends concerning employment, but gaps in the data prompted me to develop a survey that would indicate where PhDs are employed at a more granular level. In this chapter we explore the results of this large-scale survey and investigate a few representative occupational choices for scientists.

The National Science Foundation has conducted surveys of PhD recipients and postdoctoral scholars for years via different survey instruments, some of which date back to 1957. Within the NSF, the National Center for Science and Engineering Statistics (NCSES) generates statistical data for the following three areas: the science and engineering workforce; the condition and progress of STEM education in the United States; and US competitiveness in science, engineering, technology, and research and development (R&D).[1] The NSF data sets are robust and comprehensive but still miss critical research questions I was interested in pursuing.[2]

In addition to the NSF, there have been many other data-gathering projects to illustrate the employment picture for PhD-trained scientists. One of the largest ever conducted was the Sigma Xi Postdoc Survey in 2003–2004. The survey respondents included 7,600 postdocs, and the data collected encompassed demographics, career plans, satisfaction with the overall postdoc experience, mentorship, benefits, and salary.[3] Although this study was the most comprehensive ever run in terms of postdoc data collection, it demonstrated only where postdocs were employed, rather than showing where all PhD holders in the United States were employed at that point in time. Further, the study was cross-sectional rather than longitudinal, so the data captured are no longer current.

Other studies on the PhD population have been directed by scientific societies, publications, professional associations, and individual researchers, but none have endeavored to measure career outcomes of recent PhDs across disciplines that include the physical, social, computational, and life sciences. Few have attempted to capture data on skills developed over the course of doctoral and postdoctoral training and to determine their need in different occupations.[4]

Construction of a New Survey of PhD Careers

Given the gaps in data available, I decided to create a new instrument to capture the information PhDs had been seeking. This survey included questions about educational background, postdoctoral training (if any), skill development, volunteer or employment experiences outside of their research, and career outcomes.[5] The research questions addressed by this instrument included:

- What skills, if any, are developed organically during graduate and postdoctoral training?
- Are these same skills required for success in particular occupations?[6]
- In what sectors are recent science PhDs currently employed?
- What types of experiences are required for these positions?
- Are PhDs in science satisfied in their work?

This last question was of particular interest to me, since I have met with many PhDs who were convinced that satisfaction at work could only be found through a tenure-track position.

Measuring Career Satisfaction

What do the data reveal about career satisfaction in tenure-track, non-tenure-track, or nonfaculty positions?

As shown in Figure 5.1, tenure-track (85 percent) and tenured (84 percent) faculty are only marginally more satisfied in their jobs than those not in faculty jobs at all (81 percent). What's more, a greater number of those in nonfaculty positions (41 percent) indicate feeling "Very Satisfied" in their positions, the most positive category, than those in tenure-track faculty positions (38 percent).

However, we can see that far fewer of those employed in non-tenure-track positions describe themselves as "Very Satisfied" with their jobs. If we expand this analysis to include those in both faculty and nonfaculty positions across sectors, we can see in Figure 5.2 that the level of satisfaction, as measured by those respondents

How satisfied are you in your current position?

		Very dissatisfied	Dissatisfied	Neither dissatisfied nor satisfied	Satisfied	Very satisfied
If you are currently employed in a faculty position, please indicate your status:	*Tenure track*	2%	6%	8%	47%	38%
	Tenured	3%	7%	7%	46%	38%
	Non–tenure track	4%	12%	14%	45%	26%
	Not employed in a faculty position	2%	6%	10%	40%	41%
	Total	3%	7%	10%	42%	38%

5.1. Satisfaction by career field (*N* = 3,335)

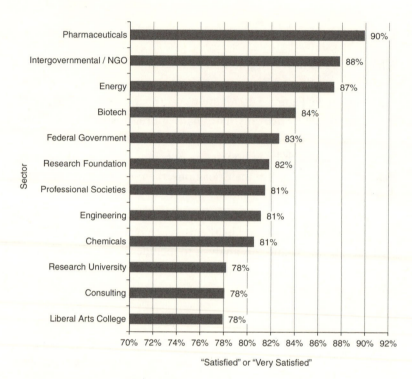

5.2. Level of satisfaction by sector

who rated themselves as "Satisfied" or "Very Satisfied," remains close to the percentages represented in Figure 5.1.

It is important to note that moving into a non-tenure-track faculty position may reduce your job satisfaction. If you presume that your logical next career step is to secure a tenure-track position, you may benefit from the self-assessment suggested earlier in this book, as well as some additional career exploration, suggested in this chapter and the next. The exercises provided can assist you in confirming your career goals. The evidence suggests that if you do seek a faculty position, you should seek tenure-track over non-tenure-track jobs to maximize your job satisfaction, but you should also develop a backup plan outside of the tenure track that interests or excites you. This plan may help you to avoid feeling "stuck" in rotating adjunct positions that are not a good fit for your needs and do not meet your expectations of a faculty role.[7]

Survey Demographics

To be eligible for the survey, respondents must have received any doctoral degree (PhD, PsyD, DrPH, ScD, or MD) between 2004 and 2014 in any of the physical, life, social, engineering, or computational sciences, and must have studied, trained, or worked in the United States.

The survey yielded 8,099 usable responses, of which 3,253 identified as current postdocs and 3,546 as currently employed PhDs in the workforce. Of the respondents, 43 percent identified as male, 56 percent as female, and approximately 1 percent identified as transgender or preferred not to respond. Ten percent identified as one of the following underrepresented groups in science: American Indian/ Alaska Native, Black/African American, Hispanic/Latino, or Native Hawaiian/Pacific Islander. US citizens or permanent residents made up 82 percent of the respondents.

Of those who had completed a PhD or PhD and postdoc (4,028), 22 percent were in tenure-track faculty positions in research institutions, either prior to going up for tenure or tenured. An additional 13 percent reported holding non-tenure-track faculty positions in college or university settings.

Because this is a self-selected group and the sample may be biased toward those in tenure-track positions, I contacted the National Science Foundation for the most recent data on PhDs who hold tenure-track positions five years out of their programs. While my sample encompasses all sciences and includes PhDs earned over the past ten years, the current data from the NSF indicate that, of recent doctorates trained in the United States in the life sciences, *only 7 percent were in tenure-track positions* five years out from their degree programs.[8]

Our survey data vary from the NSF data in part because the NSF data represent a larger sample size and a higher response rate. Given that our sample may be biased in favor of those respondents in tenure-track jobs, keeping your the career options open is critical, for the number of candidates securing tenure-track jobs may remain flat or continue to decrease.

As most candidates interested in science careers are familiar with tenure-track faculty jobs, it might be helpful to consider the remaining 78 percent of respondents to the study to determine what types of jobs PhDs hold outside of the professoriate.

Where Are PhDs Employed?

Figure 5.3 includes the four broad sectors that employ PhDs currently. Although it is interesting to note the high percentage of PhDs employed in education, it is critical to take a closer look at the types of jobs held by PhDs to get a true flavor for the diversity of occupations among this highly trained group.

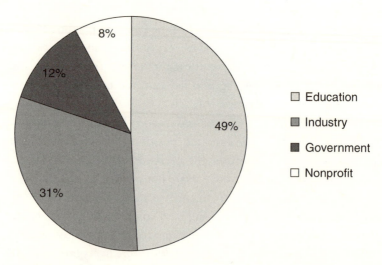

5.3. PhDs employed across all sectors

For example, here are some intriguing job titles given by participants in the study:

Volcanologist
Game Designer
Coordinator of Freshwater Turtle and Tortoise
    Conservation
Director of Institutional Effectiveness
Zoo Nutritionist
Aerospace Physiologist
Nanofossil Biostratigrapher
Principle Behavioral Psychologist
Community Nutrition Education Program Specialist
Geneticist
Virtual Lab Manager
Foreign Affairs Officer
Director, Biofuel Strategy
Coastal Landscape Adaption Coordinator
Health Informatics Innovations Analyst

These PhDs are employed in a wide range of environments, including
nonprofit research institutes, government agencies, universities, and
private corporations.

If we take a closer look at each sector, we learn more about the
particular employers of PhDs.

Jobs in Education

Nearly 50 percent of all PhDs in the study who are working full time
(3,359) are employed in the education sector. Of these, 66 percent
currently work in research universities, 23 percent in liberal arts colleges, 5 percent in community colleges, 4 percent in K–12 schools,
and the remainder in medical schools or comprehensive regional universities (Figure 5.4).

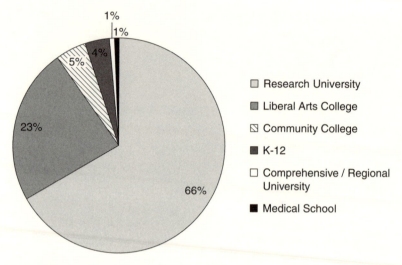

Legend:
- ☐ Research University
- ▨ Liberal Arts College
- ⬚ Community College
- ■ K-12
- ☐ Comprehensive / Regional University
- ■ Medical School

Pie chart values: 66%, 23%, 5%, 4%, 1%, 1%

5.4. PhDs employed across the education sector

In addition to the PhDs at research institutes who work in tenure-track faculty positions, there are myriad jobs across the education sector, including:

Academic Advisor
Director, Core Facility
Biostatistician
Grants Administrator
Data Analyst
Laboratory Manager
Technology Transfer Specialist
Associate Dean
Research Scientist
Curriculum Developer
Clinical Trials Coordinator
Imaging Specialist
Public Affairs Officer
Department Chair

Clinical Psychologist
Collections Manager

How might this information impact your career search? Think carefully about the kind of environment you might prefer. Do you enjoy being on a college campus? If so, have you considered exploring career options within a college or university setting? Look closely at your career development profile in Chapter 3. Based on the interests, skills, and values listed there, consider whether a career in education might be a good fit for you. As the diverse job titles suggest, there are myriad job options within this sector. Whether you enjoy working one-on-one with students, researching new technologies and bringing them to the marketplace, analyzing data, managing grants, or other activities, a career in the education sector may be right for you. Use these titles to begin exploring options, but don't stop there. Visit the "Administrative" job listings on the website of the *Chronicle of Higher Education* and look at the listings under "Academic Affairs" and "Student Affairs." Are there any job titles that interest you? Read more, and then try to find a professional who works in a similar occupation to learn more about it.

Jobs in Government

Beyond the education sector, another group of employers of PhDs in this sample are government agencies, either at the state, local, or federal level, representing 12 percent of the entire sample of PhDs working full time. Of these, 87 percent are employed by the federal government, 8 percent by state governments, 3 percent by the military, and 2 percent by local governments (Figure 5.5).

PhDs employed in this sector hold a variety of different jobs, including:

Research Scientist
Field Application Specialist

Astrophysicist
Epidemiologist
Grants Administrator
Chemist
Watershed Ecologist
Staff Scientist
Biologist
Consultant
Policy Analyst
Program Officer
Director, Core Facility
Science Writer
Public Outreach Specialist
Data Analyst
Clinical Trials Manager
Quality Control Specialist
Neuroscientist

As you read through this list of occupations within government
agencies, what do you notice? What struck me is the number of

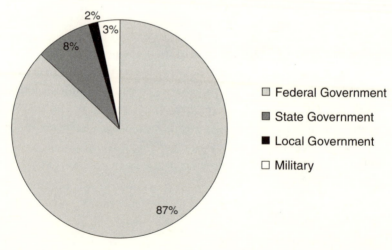

5.5. PhDs employed in government

scientists who have remained in their field of research through employment with a government agency. In addition to the list above, there were botanists, cancer biologists, environmental scientists, and others in the study sample who found work in their field of interest by exploring connections in federal, state, and local agencies.

If you are not interested in conducting research in your field but would like to stay connected to major developments in science, you might consider a career in science writing. You may want to write for those in your field, or for the lay public. To find out more, contact a science writer, a favorite blogger, or a journalist who has written a piece that you enjoyed.

Serving as director of a core facility may be attractive to you if you would like to interact with scientists on a daily basis. Consider not only your interests generally, but also your ideal job description from Chapter 1. What type of people might you like to spend time with? This question helped to inform my own career choice, as I most enjoy being around graduate students and postdocs. Asking yourself these probing questions will help you to narrow the broad field of government to focus on particular occupations that might be of interest.

Jobs in Industry

Of those PhDs employed across all industrial sectors, the majority are employed within the biotechnology, pharmaceutical, and medical device industries. These workers represent 12 percent of the overall sample of 3,359. Of these, 50 percent are employed in the biotechnology sector, 39 percent in pharmaceuticals, and 11 percent in medical devices and diagnostics (Figure 5.6).

Job titles found across the entirety of the industry sector include:

Vice President, Research and Development
Regulatory Affairs Specialist
Product Development Scientist
Medical Writer
Data Scientist

Marketing Specialist
Computational Biologist
Medical Science Liaison
Program Manager
Team Leader
Technical Support Specialist
Laboratory Manager
Technical Sales Representative
Clinical Applications Manager
Research Analyst
Business Development Analyst
Patent Attorney
Consultant
Principal Investigator

The sectors listed above—biotechnology, pharmaceuticals, and medical devices—are quite broad and represent a much larger ecosystem of life-science opportunities. Lauren Celano of Propel Careers has sketched out this ecosystem with many scientists over

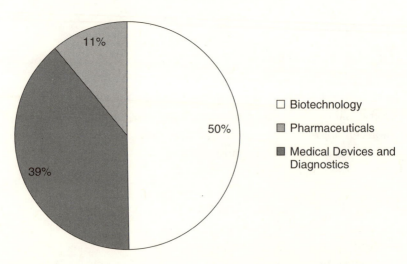

5.6. PhDs employed in pharmaceuticals, biotechnology, and medical devices

the years. Her outline includes the fields of R&D, consulting, product management, business development, marketing, clinical trials, regulatory affairs, bioinformatics, systems biology, modeling, pharmacoeconomics, reimbursement, patient advocacy, market access, finance, legal, and operations—to name a few.[9] Your research into this ecosystem may benefit from a close reading of *Career Opportunities in Biotechnology and Drug Development* by Toby Freedman. Freedman provides an excellent overview of the wide range of occupations within this life sciences system.[10]

Outside of the life sciences, there are multiple career fields in the private sector that one might explore. Figure 5.7 includes the *raw* numbers of PhDs who are employed in additional sectors.

Jobs in Nonprofits

The final cohort, representing 6 percent of the entire sample, works across nonprofit organizations, including research foundations, professional societies, nongovernmental organizations, educational services, research institutes, museums, zoos and aquaria, and botanical gardens (Figure 5.8).

Job titles here include:

Director of Bioinspiration
Policy Analyst
Editor
Director, Postdoctoral Affairs Office
Senior Scientist
Statistician
Project Manager
Science Writer
Engineer
Museum Educator
Exhibit Content Designer
Grants Administrator
Computational Scientist

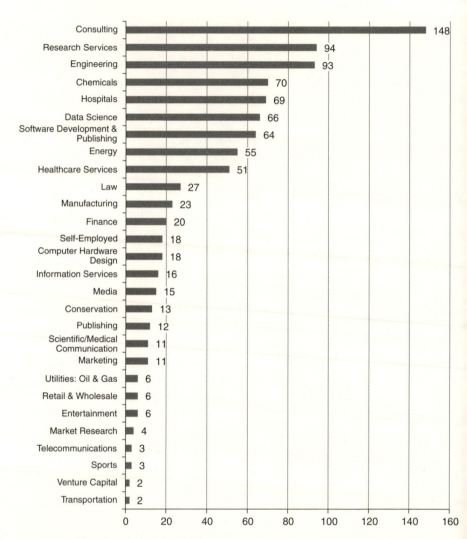

5.7. PhDs employed across the private sector

Clinical Research Associate
Public Relations Manager
Director of Nutrition Advocacy
Conservation Officer
Director, Graduate Training Initiative

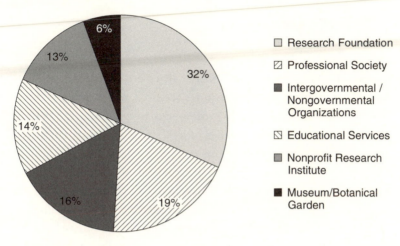

Research Foundation

Professional Society

Intergovernmental / Nongovernmental Organizations

Educational Services

Nonprofit Research Institute

Museum/Botanical Garden

5.8. PhDs employed in the nonprofit sector

If you are curious about nonprofit work, check out the mission statements of a variety of organizations that interest you. Do you feel compelled to find a cure for a particular human disease through research? Try to find a research institute or foundation whose mission fits your values. Are you interested in training the next generation of scientists in your field? Consider joining a professional society connected to your discipline. If you let your values drive your search for organizations, chances are you will uncover options that match your passion for helping others.

What Do Working Scientists Do?

Identifying sectors and occupations of interest is a worthwhile exercise, but it is equally important to think carefully about what you might like to *do* while at work. Which activities would you find most engaging? The following section outlines the primary activities of PhDs who are employed full time:

- 40 percent of those engaged in full-time work beyond a postdoc conduct basic research in their jobs;

- 36 percent teach; and
- 34 percent conduct applied research.

What do these data tell us? Nearly three out of every four PhDs are finding employment that requires research as a primary activity. Bear in mind that this research may or may not be in the field in which these PhDs trained. Further, more than one-third are employed in positions that require teaching at some level, such as teaching at a liberal arts college, informal instruction at science museums, curriculum development at the high school or college level, outreach for a foundation focused on increasing students in STEM, and more.

Other activities that PhDs are engaged in include those shown in Figure 5.9. Clearly, PhDs are employed in positions that require a great deal of task diversity. Now that we have a better understanding of the activities that employ PhDs, let's take a closer look at what may be required to obtain these positions.

Experience Required

What types of experience have currently employed PhDs engaged in before securing a job? This research question was an important one in my study, as I have worked with many PhDs who are worried about their lack of experience when embarking on the job market. Figure 5.10 may show you that you share many types of experiences with other PhDs that will be useful to you when the time comes to look for a job.

It is important to note that survey respondents interpreted "developed experience in the field" to mean experience prior to entering a paid job. This type of experience can be gained in the form of university or institutional service, volunteerism, industrial collaborations, supervision and mentoring of students, and indeed research required for their doctoral work or postdoc. For PhDs on the job market, then, it is crucial to translate experiences you have already had into concrete experiences on your resume or CV to convince an employer that you are a viable candidate.

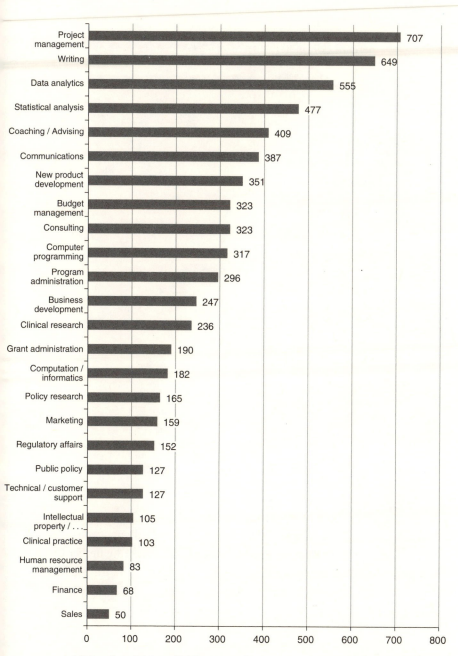

5.9. Additional activities of PhDs at work

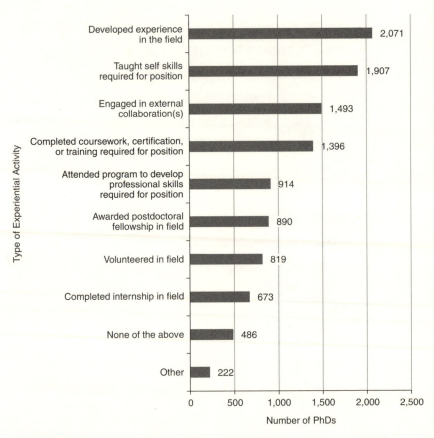

5.10. Experiential activities of PhDs during their postdoc or course of study. "Other" = campus or student organizations, networking, additional degrees, informational interviews, or job shadow.

In addition to listing experiences that they had engaged in prior to their current position, respondents were also asked whether these had been helpful in obtaining their current position. Of 3,206 respondents, 85 percent indicated that these experiences were helpful in getting their current jobs. Chapter 6 presents concrete steps you can take to gain skills and experience in your field of interest, including the strategies listed in Figure 5.10 and more.

Federal agencies are also supporting PhD skill development and hands-on experiential activities through funding mechanisms, including the NIH's awards, Broadening Experiences in Scientific Training (BEST). In 2013–2014, these awards were distributed to seventeen research institutes across the country to develop training programs for PhDs in science "that reflect the range of career options that trainees may ultimately pursue and that are required for a robust biomedical, behavioral, social and clinical research enterprise." Each of the seventeen research sites includes a website with more information about career options, as well as strategies for building experience and skills in particular areas.[11]

For further examples of the types of experience PhDs bring to different occupations, let's look at a few PhDs currently employed in a variety of professions. All share their understanding of skills and experiences needed to move into these fields.

Snapshots of Currently Employed PhDs

The following six professionals volunteered their time to talk with me about their jobs, skills required, and strategies for building those skills. Each holds a PhD in one of the physical, life, or social science disciplines, has completed postdoctoral training, and is quite satisfied in his or her current occupation.

Raluca Ellis

PhD, Chemistry, University of Toronto
Climate and Urban Systems Partnership (CUSP) Director, The Franklin Institute

Raluca Ellis plans, manages, and oversees the day-to-day activities of the Climate Change Education Partnership, designed to improve climate literacy in urban settings. The multicity project is a collaboration of four science museums in Philadelphia, New York,

Pittsburgh, and Washington, DC, each of which will act as a hub to develop and implement the CUSP model in their home cities.

Ellis maintains and supports this collaboration to ensure constant communication among the partners, sets and monitors project goals and milestones, and coordinates all aspects of the project, including timelines, budgets, and required reporting systems. She is responsible for the planning and oversight of Philadelphia's Urban Learning Network, making sure that local partners participate and are fully engaged.

In addition to her CUSP duties, as an expert in Environmental Science, Ellis supports the Franklin Institute's environmental exhibits and programs. Most of her work takes place in an office at a desk. But occasionally she interacts with visitors on the museum floor or engages with residents in the community.

Key skills required by this position include:

- project management;
- time management;
- people management;
- facilitation skills;
- communication skills (written, verbal, interpersonal);
- teamwork;
- creative problem solving, and
- science presentation skills—the ability to take a complex topic or issue and distill it to bite-sized, conversational language.

On how she built the skills required for this job, Ellis notes: "I think you gain most of these skills during your PhD and postdoc years, especially the project and time management since we usually have a million things to do at once . . . Communication skills can be built from doing science outreach with a variety of audiences." Such skills could be developed by volunteering at a local science center or volunteering at a science festival. Practice writing, blogging, and explaining what you do to a lay audience. Finally, Ellis mentions, do not forget to network: meet with those who work in similar institutions and ask them about the daily joys and challenges of their jobs.

According to Ellis, the average salary range for PhDs entering this field is between $55,000 and $65,000. "It's a nonprofit museum so you're not making tons of money. It's the mission that counts." She concludes, "Science communication is being able to take something you are NOT an expert in but still know more about than the average person, and still make sure it is scientifically sound—in 10 seconds."

Alok Tayi

PhD, Materials Science, Northwestern University
Entrepreneur, Cofounder, PreScouter

Alok Tayi emphasizes that "the only thing 'typical' about being an entrepreneur is that you are always putting out fires." A self-made job such as Tayi's may involve product development, R&D, sales, finance, and marketing. The environment may also be much more flexible than for other positions: "starting a company occurs everywhere! It can be in an office or in a workshop or virtual," Tayi notes.

Being in charge of an enterprise requires skills that may sound familiar to you from your work as a PhD or a postdoc. You must have a deep and thorough understanding of a problem, be an able leader, and communicate well with others. Having a strong stomach for risk is also a requirement.

Tayi proposes a number of concrete strategies for enhancing your skills in these areas. These include gaining experience in contract research organizations, which provide flexibility and an introduction to industry needs. Getting to know people with technical skills in business, economics, and finance will enlarge your understanding, and sales experience will help you learn how to listen, communicate effectively, identify needs, and build trust. In addition, consider taking an abridged MBA course, if possible. Tayi emphasizes the importance of learning how to code: "This is an important skill and will continue to be a need that permeates all sectors." Look carefully at skills you may already possess that cross boundaries. These assets may help you develop interdisciplinary collaborations. Finally, Tayi says, "One can do some of these things independently via work experience or personal passion. However, the risk piece never becomes easy."

In such a high-risk occupation, it's hard to predict a typical salary, says Tayi. It may vary depending on what stage your business has achieved or what level of funding you have obtained. You might make as much as $80,000 a year or nothing at all.

Holly Dail

PhD, Physical Oceanography, Massachusetts Institute of Technology/Woods Hole Oceanographic Institution
Senior Quantitative Researcher, The Climate Corporation

Holly Dail's job may sound similar to that of a typical scientist. It includes technical work, data analysis, code writing, and other elements of research. She works in an office, but also telecommutes from home. Dail, however, has made a significant change in her career. She now works in atmospheric science, even though oceanography was her field of study.

The skills required for her job may sound familiar, too. She lists as essential strong analytical, quantitative, and verbal skills in addition to the skills in written communication typically acquired during the PhD process. Personal skills play a factor, too: flexibility, adaptability, willingness to learn new things, and the ability to work with diverse groups of people are assets in such a collaborative environment.

Dail emphasizes the strategies that allowed her to refocus her research, using her acquired skills in a different academic discipline. Take a programming or statistics course for practical skills that transcend any particular discipline. Remain open to being stretched intellectually. Go to seminars in disciplines outside of your field, and express an interest to others in learning about their discipline. Your energy and interest in others' work will demonstrate your flexibility.

The average salary range for such a research position is hard to predict, according to Dail, as it is dependent on your experience level and the employer.

Steve Bennett

PhD, Theoretical Physics, McGill University
Lead Data Scientist, Rovi Corporation

Steve Bennett leads a team of data scientists and analysts who measure and predict audience behavior in the entertainment industry. He and his team develop algorithms leveraging large sets of structured and unstructured data from multiple sources, applying techniques of machine learning, simulation, and nonlinear optimization. They are responsible for sharing their research with software developers and customers. He works in an open-concept office in a major city.

Bennett suggests that an ability to turn "fuzzy ideas into something concrete" is an asset in his position. More concrete skills include:

- the ability to code in Python or R;
- knowledge of machine learning;
- familiarity with databases;
- "soft" skills; and
- the ability to explain the big picture of your research.

Bennett suggests some useful strategies for building these skills. Take your favorite paper, he suggests, and explain it to a friend. Describe why it matters, what problem you solved, and what was new about it. In interviews, never "give a science talk." Develop a side project that involves data and demonstrates your skills. Enter predictive modeling competitions such as Kaggle to test yourself. Meet people who share your interests through meetup.org or similar platforms. Finally, do not neglect the practical: take online courses through Coursera or others to learn a general programming language, such as Python.

The average salary range for PhDs starting out in data science, according to Bennett, can range from $90,000 to $130,000.

Rohan Manohar

PhD, Pathology, University of Pittsburgh School of Medicine
Marketing Specialist, Clontech

Rohan Manohar works with a product manager on products that his corporation, Clontech, is developing, specifically CRISPR/Cas9 products. Manohar, who works in a typical office environment, spends one-third of his time involved in sales representative training and support. This involves creating training videos, cue cards, and product launch training programs, as well as developing sales calls and creating sales collateral for seminars and vendor shows. The other two-thirds of the position mixes product and project management. Product management entails: working with a boss to think about the market positioning for a product and how or if it addresses a customer pain point; communicating marketing messages via email campaigns, landing pages, and application notes; and working on search engine optimization with Google remarketing ads and Adwords. Project management includes streamlining the development of ongoing projects to fit the specifications and market requirements, evaluating which new projects the market needs or would support.

Manohar suggests three skills as essential to the position: business acumen, understanding the "big picture," and the ability to distill the technical complexity of a project down to one or two lines that convey its utility, usually in terms of human health.

How to acquire these skills? Manohar recommends taking business classes, networking with people in business, volunteering, or interning with a business if possible. Manohar recalls an internship in business development at Woodland Pharmaceuticals. A marketing position requires a person who can think about the "market needs" of a project and how to sell it.

The salary range for a PhD entering a marketing position similar to Manohar's might range from $70,000 to $85,000.

Jim Gould

PhD, Biochemistry and Molecular Biology, University of Louisville School of Medicine
Director, Office for Postdoctoral Fellows, Harvard Medical School

Jim Gould works in a university environment—mostly in an office except for time spent attending or delivering workshops. Some of his time is spent in typical office tasks. He troubleshoots postdoc concerns, answering phones, triaging visitors, and answering questions via emails. He manages the listserv and the office budget, and creates the marketing materials used. But he must also think strategically about his office's function, developing workshops, seminars, and other aspects of career coaching programs offered. He must consider how these programs fit into a cohesive curriculum.

Gould's list of skills that his job requires includes:

- enthusiasm;
- energy;
- office management skills;
- budget management skills;
- programming skills;
- the ability to understand local and national policy; and
- career coaching skills.

Gould suggests saying "yes" to volunteering opportunities. By working behind the scenes in an administrative capacity, you may learn that administrators can be allies. Is starting a seminar series of interest to you? Start doing what you're interested in doing, he says, concluding, "Don't put off until tomorrow what you can do today!"

Starting salaries for PhDs entering this field vary greatly by experience and institution, but Gould estimates a range of $60,000 to $100,000.

Manisha Sinha

PhD, Interdisciplinary Graduate Program, Graduate School of Biomedical Sciences, University of Massachusetts Medical School
Scientist, Drug Development, Biogen Idec

Manisha Sinha works principally in the lab; she estimates only about 30 percent of her time is spent in an office. Sinha works in cell biology with, among others, proteomic scientists and biochemists. Whatever their background, everyone is applying different approaches to the same problem. Most of the nine-to-five day is spent at the lab bench, with time allotted for conducting data analysis and writing up results. Although the position offers the flexibility to telecommute and plan one's own time, Sinha needs to be in the office for experiments and meetings.

Sinha lists collaboration as a key job skill, with its attendant skills of receptivity, listening to others, and respect for others' opinions. A solid publication record and understanding of one's field are necessary. She emphasizes the importance of publishing, starting collaborations with others, and finding models for your work among different PIs.

Salaries for entry-level drug development scientists can be very competitive—beginning at $70,000 to more than $100,000, depending on your experience level and your employer.

Moving Beyond the Data

The career options shared throughout this chapter represent a dynamic, not a static set of possibilities, as new fields are emerging daily. Given the proliferation of start-ups, new technologies, and new job titles that didn't exist in the past ten or even five years, we will continue to see the scientific workforce expand in unexpected directions.

Your next step on the career development path is to explore any of the careers mentioned here, or by a colleague or friend, or found in print or online. Chapter 6 will provide tips on how to explore any fields of interest. Take advantage of the opportunity to explore who you are and what you might like to do. Learn to match your skills with the right job for you. In Part Three of this book, you will learn how to present these skills in ways that showcase how your expertise and skills are relevant for the positions you are interested in applying for. In the interim, work to gain experience in your fields of interest to improve your skill set, to show your commitment to the job area, and to gain connections for networking and references that will be relevant when you are ready to apply for jobs.

There are many exciting jobs out there for PhDs. Commit to learning more about opportunities that interest you, and be willing to reassess your career goals as you learn more about different career opportunities. Career development is a continual cycle of reflection and re-evaluation.

CHAPTER 6

Strategies for Exploring Careers and Building Experience

Conducting occupational research will help you to decide whether different types of jobs might be a good fit for you. In this chapter I discuss a number of different approaches you can use to learn more about various career fields—and how to gain hands-on experience in those fields. Some of this research can be conducted online, such as reading job descriptions or visiting professional association websites. Other methods require deeper involvement: conducting informational interviews; volunteering; collaborating on a project; interning or becoming a fellow; taking classes; shadowing a professional, and—at all times—networking! We will investigate each of these methods in turn.

Read Job Descriptions

To understand what a job requires both in terms of skills and experiences required for entry *and* tasks and responsibilities required on the job, one place to start is with an online job description. You can find sample job descriptions just as easily as you can find definitions for various fields.

If you were to search for "technology transfer," you might find something like this:

Technology Transfer Associate

Office of Technology Development

The Office of Technology Development (OTD) is looking for the right person to handle responsibilities for an ever-expanding portfolio of innovations/intellectual properties, including patents and copyrights. The Technology Transfer Associate (TTA) will identify new technologies, work with innovators to chart the course from bench to the marketplace, develop partnerships with industry and investors, negotiate option and license agreements, catalyze collaborations with companies, and much more. The TTA is expected to have service-focused interactions with faculty. He/she will assess intellectual property and marketing landscapes of new innovations. The TTA will write marketing pieces for the OTD website, develop brochures, and engage in other marketing activities. The TTA will help faculty with material transfer agreements and other related efforts. The position will include but not be limited to the following functions:

- Actively market university innovations to the private sector.
- Draft and negotiate intellectual property options and license agreements.
- Perform intellectual property mining, analysis, and marketing landscapes.
- Develop marketing pieces for the OTD website, develop brochures, and generate other materials in support of OTD's marketing efforts.
- Provide consultative assistance to faculty, staff, and students around intellectual property issues.
- Foster and facilitate entrepreneurial activities.
- Provide broad-based intellectual property support in such areas as basic sponsored research, clinical sponsored research, material transfer agreements, confidential disclosure agreements, inter-institutional agreements, etc.
- Assist the Director and Associate Vice President as needed.

Job Requirements

Minimum Qualifications

This position requires an advanced degree in the life sciences with at least 2–3 years of directly related and progressively responsible work experience.

Preferred Qualifications

A strong preference will exist for candidates trained at the doctoral level who also possess demonstrated experience in transactional work such as material transfer agreements, confidential disclosure agreements, inter-institutional agreements, and options/licenses. In addition, a keen understanding of intellectual property law, particularly with regard to patents and copyrights, is essential.[1]

This description contains enough information to give the reader a solid understanding of the position, as well as its requirements. This also provides you, the applicant, with two other important pieces of information: suggestions for skill development in areas where you might be lacking, and language you can use to strengthen your application materials when you are ready to hit the market for this kind of job. This is a particularly useful example. Other job listings will not be as descriptive, but continuing with your search of job listings will help you find both comprehensive descriptions and multiple listings so you can compare and contrast different positions. For example, in my search, I came up with several different job titles. Having a list of possible keywords for job titles will help you know what to look for if you were to start searching for these kinds of jobs:

Technology Development Specialists
Licensing Associate
Intellectual Property Manager
Technology Commercialization Manager

In addition to different titles, I can compare the responsibilities and requirements of this job by type of organization. Does the job vary if it is located in a university versus a national lab? What if it is based in a pharmaceutical company? Studying descriptions will help you to answer these questions, and in the process, strengthen your understanding and determine your level of interest in a given field.

Visit Professional Association Websites

As it happens, a search on "technology transfer" led to the website of a well-established professional association with a career center webpage. In this case, it was the Association of University Technology Managers (autm.net). Many professional associations have a robust career-center link on their website, as the mission of a professional association tends to include the growth of the profession.

Whether listed under a "career center" heading or another link, professional association websites include detailed descriptions of their professions, as well as the training and education required to enter them. A great example of this is the site of the Regulatory Affairs Professionals Society (www.raps.org). This site lists books, e-courses, and other learning avenues for PhDs interested in becoming regulatory affairs professionals.

Some PhDs have expressed concern that, with so many existing occupations, there cannot possibly be a professional association tied to each. However, it is surprisingly simple to find an organization tied to any occupation you can think of. For example, a search for "science writers association" leads to the National Association of Science Writers (www.nasw.org). "High school teachers of science association" will bring you to the National Science Teachers Association, and so forth.

Find—and Attend—Local Meetings

Once you have located the website for a professional association of choice, there are several sections that will be of use to you as you continue through career exploration and an eventual job search. One of the sections I encourage you to visit is the "events" section of the site. In this section, you will find meetings, seminars, networking socials, and other events held by the association in question. (Note that this kind of information is sometimes found under a link called "About Us," followed by "state" or "regional" association information.) Attend meetings held by professional associations if at all possible, because they present a tremendous opportunity for you to learn more about the profession *and* to network with professionals.

Contact Board Members

Another area of professional association websites to explore is the Board of Directors link. You may find this link on the front page, or under "About," or "Leadership." The list of professionals you will find on such a page represents an unprecedented opportunity for you to connect with leaders in your field of interest. These pages often include pictures of board members and almost always include email addresses. As you seek to learn about different professions, board members represent an easy group to approach, because they are all volunteers, offering their time and expertise to others in the profession, and they are indeed *charged* with building the profession. Consider this a ready-made group of experts for you to contact, as most will be delighted to field questions from you about their profession and will be willing to give you their time.

Conduct Informational Interviews

Once you have identified professionals in a given association or society, through a local meeting, through a friend, or on campus at an employer or networking event, schedule an informational interview with that person. Informational interviews can be defined quite simply as the process of gathering information on occupations by interviewing professionals in those fields.

A CREATIVE WAY TO LEARN MORE ABOUT BIG DATA

I worked closely with a postdoc in physics who was interested in big data. He had applied for a training and internship program in data science to learn more. He was a finalist but did not make the final round of cuts. Recognizing that he needed to develop his skills further, he came to the Office of Postdoctoral Affairs to learn more about the field.

Together, we found that a huge organization was hosting a conference on big data in Boston in the next few months. As the conference registration fee was quite high, I told him that the postdoc office could offer him sponsorship to the meeting if he offered to teach a workshop for graduate students and postdocs on all that he learned from it, including what the field entailed, how and where it was growing, and which employers were hiring.

At the conference, he made several important contacts, learned a great deal about the field—including where he needed to strengthen his skills—and upon his return, presented a workshop to a packed room of graduate students and postdocs interested in the same.

Based on this experience and his subsequent efforts to build skills and follow up with employers, he secured a job in data science within months.

Why conduct informational interviews? The best source of information about an occupation is most often a person working in that occupation. When planning for informational interviews, I recommend engaging not one, but several professionals in conversations about their work, since you may catch someone on a bad day and hear nothing but negative thoughts about their work.

How to Request an Informational Interview

When first reaching out to strangers to request an informational interview, be clear about your purpose: this contact is *not* meant to be used as an opportunity to ask for a job. This contact *is* meant to be your chance to learn more about a profession, and about a particular person's path.

Reaching out via email is the least intrusive way to get in touch with people. Try using a message like this:

Hello, Dr. Higgins!

My name is Sandy Duncan, and I am currently working as a postdoctoral researcher at the Fred Hutchinson Cancer Research Center in Seattle. My focus is molecular biology, and our research group is studying the mechanisms that control the process of blood vessel formation.

I am writing to you now because I am interested in making a transition into science writing. I have served as editor of our Postdoctoral Association's bimonthly newsletter for the past year and am now exploring writing and editing as a full-time occupation.

Might you have time to talk on the phone for 30 minutes in the next few weeks? I would love to hear more about your work at *Nature Methods* and about your career path more generally.

Thank you for letting me know what times and days work best for you!

All the best,

Sandy

Many graduate students and postdocs have shared their reluctance with me to reach out to strangers in this way, but in my experience, in most cases professionals that you contact will be more than happy to speak with you about their jobs. In fact, many may be quite flattered, since you are indicating through your query that you consider them to be experts in their own field. I have sat with countless PhDs who have contacted professionals and expressed shock at the results—shock at the kind and gracious response they encountered, even from the most accomplished researchers in a given field!

I have been on both ends of this conversation. I have asked experts in my field for more information about their work and have always been pleasantly surprised by their generosity and kindness. I have also been contacted by many PhDs over the years who expressed an interest in my work in career counseling and university administration, and who requested an informational interview with me. Each time, I was truly happy to share what insights and perspectives I could with these researchers, as many professionals have helped me over the years.

Professionals from all occupations are typically delighted to hear from and engage with junior researchers.

Do not hesitate to contact people using the form invitation given above. Chances are great that you will find people who are happy to talk with you about their work.

Identifying Professionals to Interview

Another way to identify professionals for informational interviews is through a literature search. If you are exploring careers in biotech, for example, and want to speak with a staff scientist working in research and development, search for the name of a company you are interested in joining in a publication database like PubMed. As you know, articles that appear in peer-reviewed journals include the personal emails of the authors, so you can identify and reach out to industrial researchers this way. In this case, a logical opening point of conversation is the paper in question, and you should include this information when emailing the researcher.

LinkedIn can also be a valuable source of information about professionals in different fields. If you are interested in museum work, you may search for the name of a local science museum to find a list of professionals who work at that site, and then read through their LinkedIn profiles, if visible. If the names are visible but the profiles are not, conduct a web search for that person to learn more about their background before reaching out to them.

If you are contacting professionals who are local, try to set up an in-person meeting. Face-to-face meetings are much more powerful than telephone conversations, and you are more likely to be remembered. Be aware, however, that if you invite a local professional to have coffee or lunch with you, *you* need to pick up the bill.

Having identified professionals who interest you and arranged a time to talk, you must now prepare for the meeting. Draft some specific questions for your contact about his or her work. If you are not sure where to start, try using these questions during your meeting:

Career Planning

How did you decide on your career field, and how did you get into the field?

What specific skills (for example, organizational, analytical, or interpersonal) and personal qualities are necessary in your job?

Job Search

How do people find jobs in your field? Is any method more effective than another?

What should graduate students and postdocs stress on their CVs/resumes and during interviews to get a job in your field?

Description of Career Field

Tell me what you are responsible for, and what an average work week/day involves.

What sort of career path is common in your field?

What does it take to be good at your job?

What are the sources of satisfaction in your work? Frustration?

What are the most frequently occurring and difficult problems?

What type of training is provided? What are starting salaries like?

What are the criteria upon which performance is evaluated?

What are some of the more pressing issues and concerns facing people in your profession?

You may have your own questions to add to this list, or you may just follow where the conversation leads.

Take notes during the meeting if you find that helpful, and always close by asking whether there is someone else your contact would recommend you get in touch with. This will leave you with yet another contact in the field, and potential interview subject.

Request a business card before you depart (if meeting in person), and always thank professionals for their time. Immediately upon returning to your office, desk, or lab, send an email thanking your interviewee for giving up time and offering insights to you as you continue on in your search. This is common protocol and will allow you an opening to reiterate your interest and to stay in touch with that contact.

Volunteer

Another way to gain greater knowledge of a career field is through direct experience. Spending time in a particular position will give you an insider's view and will help you to determine whether a given occupation is a good fit for you. In addition, volunteering will help you develop valued and possibly needed experience.

Volunteering with an organization of interest is a wonderful way to learn about an occupation, gain experience without leaving your program or postdoc, and build your CV or resume. Another benefit of volunteering is that you can choose your hours, which will enable you to test the waters of a particular field without interfering with your current research.

To uncover volunteer opportunities in a field of interest to you, simply reach out to organizations and ask whether you can volunteer your time. Try to be specific about the skills and abilities you can bring to their organization, and how you can contribute to their mission.

Collaborate on a Project

Many scientists and engineers I have met have expressed an interest in working in industrial science. Some have a greater interest in

VOLUNTEER TO BUILD EXPERIENCE
AND A SENSE OF PURPOSE

Raluca Ellis was a postdoc at Harvard University who was interested in museum work—not in museum education, per se, but in exhibit content development. Because she was an atmospheric scientist working on a data project, she had not obtained any concrete experience in developing content for museum exhibits but had learned about this career through her research on museum job websites.

As she read through job descriptions of exhibit content developers, she came to realize that what she was lacking was not the *scientific knowledge* needed to develop content for museum exhibits, but *concrete experience* doing that work. She decided to volunteer locally for the Museum of Science in Boston. Although she worked with young children in the Discovery Center and not on exhibit development, she was still able to learn more about how a museum operates and the kinds of positions available. Most important, she met with exhibit content developers and found out more about their work. Because she was a volunteer, Raluca was able to choose hours that did not conflict with her research schedule.

At another point in her time at Harvard, she volunteered for the Cambridge Science Festival, a huge event bringing local scientists and science to the public. In terms of this work, Raluca shared, "My interest in outreach was taking off. I was putting science out there for people, and you get to see it immediately. People were being exposed to things they had not thought of before, and it was inspirational."

Finally, she took a course in museum exhibit content development at Harvard Extension School. In that course, she had to plan and develop an exhibit, and she chose to focus her project on climate change. In constructing her ideas, she contacted and interviewed several exhibit developers who had created exhibits on climate change in their own museums, thus building a strong sense of what was required for this work.

Given her volunteer experiences and her directly relevant coursework, Raluca applied for and was offered a position as Director of the Climate and Urban Systems Partnership and Environmental Scientist at the Franklin Institute Science Museum in Philadelphia. Reflecting on her training as a scientist, she offers: "I think it's so important for early career scientists to really understand and recognize how important science communication is. This should not be an afterthought to your research—it should be as important as the primary research that you do. Communicating the work that you do to the public, and creating a more science-literate public, is crucial."

large, well-established firms while others are more curious about the start-up environment and endeavor to conduct research in that setting.

If you would like to engage with a local start-up, you might consider approaching a member of the leadership team and offering your services as a part-time consultant—perhaps analyzing data sets, setting up experiments, or identifying other tasks that align your skills and experience with the needs of the small firm you approach.

With larger firms, you might consider proposing a potential research collaboration to your PI. Many such university-industrial partnerships and collaborations exist. It is your responsibility, however, to think critically about the needs of an organization, the research they are conducting, and how your skills and experience might be a fit for what they are working on. You will also need to think carefully about how work with an industrial firm might fit into your PI's research plan, identifying the benefits to your current research group (other than building networking contacts for you).

Once a collaboration has been established, the rest is up to you. You will need to shine in your work and in your interactions with the industrial research group, as these scientists will not only become networking contacts for you but may even serve as references down the road. Take on a leadership role in the project if you can, and do your best work. Offer to be the point person on the project so that you can communicate directly with the members on the industrial research team. This experience will not only strengthen your CV for industrial research jobs, but will provide an insider look into working in a corporate environment.

Intern or Become a Fellow

Similar both to volunteering and establishing collaborations, interning with an organization will help you to develop concrete experience and skills in a given field and will allow you to get a better sense of whether that field is a good fit.

Formal Internship and Fellowship Programs

Some organizations, including some consulting firms, offer long-standing internship programs for PhDs. For example, the RAND Corporation, a research institution focused on domestic and international policy, hosts a Summer Associate program for full-time students engaged in either a doctoral or professional degree program.[2] This program, structured like other formal internships, includes a formal application process, a twelve-week onsite work period over the summer, a stipend, the opportunity to network with others at the organization or beyond, and professional development activities (varies by organization).

Another popular fellowship program for PhD students and postdocs less than two years from their degree is the Presidential Management Fellowship (PMF) Program. Historically focused on the humanities and social science disciplines, the PMF program expanded to include science, engineering, technology, and mathematics disciplines with the class of 2014.[3] The PMF program provides an opportunity for PhDs interested in research and development on the federal level, or other policy work in the United States, to spend time developing relevant skills, as well as hands-on work experience in an agency setting.

Self-Made Internship

Internships are most often—but not always—hosted by large organizations. This is true because they can be expensive to coordinate, and with recent US legislation,[4] must be offered for pay unless certain legal criteria are met. Still, it may be worth your time to approach any organization, large or small, to inquire about the possibility of an internship.

Be specific when reaching out, detailing your interest in the organization, your thoughts about how such an internship might work (hours and location, for example), your current position or research topic, and how you can benefit the organization through your experience and skills. Some PhDs I have worked with have contacted organizations directly to inquire about crafting an internship and have been successful in developing programs for themselves that included skill building and direct experience in a given field.

Take Classes

Another way to test your interest and gain experience in a field is to enroll in a course related to that field. Building skills required for entry into a particular field is critical, but may not be too time-consuming, depending on the course. You might find that you value the structure and interaction of a course at your or a nearby institution, or you might find that the pace and flexibility of an online course fits better with your other work.

I once worked closely with a postdoc who was engaged in stem cell research. The work fascinated her, but ultimately she was more interested in assisting with bringing scientific discoveries to the marketplace. Because she knew she lacked an understanding of the world of business, she enrolled in courses at her institution in financial accounting, management, and organizational behavior. These courses equipped her to analyze the market for firms seeking particular innovations.

Check with your institution to find out whether you can enroll in local or online courses. As always, you will want to honor your role in your research group, making sure that any courses or additional work outside of the classroom does not interfere with your ability to get your research done. Some institutions offer a lower tuition rate or may even waive the tuition for certain employee categories, so be sure to check your benefits carefully to see whether you can enroll at a discounted rate.

There are also many courses online that are either free or quite affordable. Most are open to anyone. Once you have identified a field of interest and recognize the gaps in your skill set, consider taking online courses at your own pace, and in blocks of time that work for your research. Building skills in this way will not only increase your value for potential employers, it will give you a better sense of whether you would be interested in applying those skills regularly in a new job.

Highly motivated PhDs may also choose to teach themselves a particular skill. Steve Bennett, a theoretical physicist, found this approach to suit him best. He sought entry into a data science position, recognized what skills he was lacking, and taught himself how to code in Python. This proved a valuable language for him to learn, as many data science employers were seeking this skill, and he was subsequently offered several positions, one of which he accepted.

Whether using books or online tutorials, there are many ways you can teach yourself a particular skill or develop experience in a given area. To remain accountable to yourself throughout this process, we will discuss how to set achievable goals in Chapter 8.

Shadow a Professional

To get a taste for what a specific workplace looks and feels like, and to understand what actually happens in a given occupation from one day to the next, you may want to consider shadowing a professional for a day. Shadowing means precisely what it implies: you will accompany that person throughout the day, following the professional to meetings, listening to phone calls, and observing interactions with staff, clients, or others.

When contacting someone to explore shadowing, be sure to make your goals clear. As with informational interviews, this is *not* an opportunity to ask for a job. This is instead an opportunity to observe a PhD in a field of interest as they go about their work day. As you reach out to people, introduce yourself first, your current program or position, and your reasons for desiring a transition into their particular career field. Following this introduction, express interest in observing their work for a day. You need not even use the word "shadow" as long as the language you use is clear.

While this request is more time-consuming for professionals than a thirty-minute informational interview, you may still find people who are willing to be your host at their work site for a day. If you don't ask, you will never know, so it's worth trying to get this bird's eye view of your desired occupation.

Network

Concluding Part Two of this book is an entire chapter devoted to this topic. Why mention it here, then? Because it bears repeating. Networking is *the* most valuable tool you can use to find a job, as you will see, but it is also valuable in learning more about a variety of fields.[5]

We have in fact discussed networking several times in this chapter already. Contacting board members at professional societies? Reaching out to people for informational interviews? Emailing employers to discuss a possible internship? Enrolling in a course, online or at an institution? Meeting people while volunteering? Building research collaborations? *All of these consist of networking.* It happens everywhere, all the time. It is pervasive, as it should be.

Keep in touch with all of the people you have contacted throughout the career exploration process. You never know when a position may arise at one of the organizations you have interacted with, and having kept in touch with professionals there, you will have a distinct, measurable advantage over other candidates interested in the same job.

Identifying Careers of Interest

Using the strategies listed above, you will undoubtedly come to a point where you can identify one or several careers of interest. In the next section, you will be approaching a career decision and drafting a plan. In order to go through that process, you will need to pinpoint a manageable number of careers to continue to explore and potentially pursue.

In your career notebook, list three careers you have explored that have piqued your interest sufficiently to consider them further. After listing each occupation, describe what draws you to each. This list will assist you as you develop your career plan and set goals in subsequent chapters.

CHAPTER 7

How to Network Effectively

Think about *networking*. The word itself can evoke powerful feelings in people: some may feel stressed out, anxious, or uncomfortable when thinking about the act of networking. Others may feel it is a superficial way of interacting with people. However, many understand the value of engaging with other people about their ideas, their work, or their goals. Networking is simply a word used to describe the variety of ways in which we connect with people—and how that process is mutually beneficial to both parties.

Fundamentally, networking is the process of *building relationships to share information.* Given this definition, all scientists have been networking since entering the field—including you. And as it happens, networking is critical both for your scientific research *and* for your career development.

If you have ever shared resources or articles with someone you met at a conference, collaborated on projects with other researchers, or spoken to someone considering your undergraduate institution, you have engaged in networking. In fact, networking is the process that keeps the research enterprise going.[1]

You may recognize the importance of networking but may not be sure how to do it. In this chapter, we will identify different types of networking, learn how to make use of contacts in your existing network, and explore strategies for approaching potential networking connections. We will discuss ways in which networking can impact your professional identity—and uncover job opportunities.

Types of Networking

The Elevator Speech

You may have heard this phrase before, and may in fact have used an elevator speech to describe your work to a stranger or a relative. An elevator speech is simply a short introduction of who you are, what you work on, and what you would like to do—a summary brief enough to start and finish on an elevator ride in a conference hotel.

I actually used an elevator speech when I met my husband. He was seated next to me on a plane and I struck up a conversation with him. That is not to suggest that you need to start flying more often to network and potentially meet a life partner. It is simply to suggest that you need to be aware of how often networking occurs organically—and be open to it and ready for it when it happens.

Try using this template to sketch out and practice your elevator speech with a colleague or friend. Practicing may seem silly or forced, but many PhDs I have worked with have struggled to explain their work very simply, in a few short sentences, when asked.

I am (who you are) focusing on (what you study). I will finish in (when you graduate or when your appointment ends) and am looking for (what you are looking for).

Here is an example:

I am a graduate student at Columbia focusing on cancer metabolism research. I will finish my PhD in May of 2016 and am looking for a research role in a biotech firm in Boston.

Sharing these key pieces of information with the person you meet will give him or her a sense of your circumstances and what you hope to do next. By providing this brief context, your new connection will be able to flip through a mental list of projects and contacts to find common ground.

Stealth Networking

A colleague of mine, Laura Stark, coined this phrase. I use it here to demonstrate that we are all networking constantly, whether we consider it networking or not. Some ways in which you have engaged in stealth networking may include attending departmental seminars, lab meetings, and journal clubs; actively participating and asking questions during talks; going to departmental social events; asking for feedback on manuscripts and presentations, and offering to critique others' work.

Networking happens every day, all around us. In addition to interactions that occur organically, I encourage you to reach out directly to people whose work you find interesting. A few years ago, I read an article in *Scientific American* about the introduction of the first female lab scientist LEGO mini figure.[2] The author, Maia Weinstock, included a link at the end of her story to a page of more than eighty real scientists and engineers, male and female, that she had depicted using LEGO mini figures.[3] I spent most of an hour looking through the images, so I decided to reach out to Maia and let her know how funny and clever I found her webpage. As it happened, she lived near me, so I invited her to lunch, and we had a wonderful discussion about women in science and changes in social media. By the end of our conversation, the topic had turned to my work on this book. Maia, a science writer and editor, offered to help in any way she could.

These types of meetings between complete strangers—professionals who take an interest in one another's work—happen *all the time*, so do not hesitate to reach out to someone whose work intrigues you. Such interactions may serve you both in your research and in your career.

Networking Online

Technology is another platform that we use constantly to network. Whether through email, social media, blogging, or another online tool, we are out there every day, conducting research, interacting with friends, and building our identity online.

Given the relative ease with which employers can find information about you, it is important to keep your online identity professional. For example, have you ever Googled yourself? Are you pleased with what you found? Are you willing to have potential employers do the same? If not, take control of your identity and make changes to ensure that your profile is consistent across all sites, platforms, and social media accounts.

LinkedIn

Q: How many employers use social media to find
candidates?

A: Across all sectors, 77 percent of employers, including academic employers, use social media to identify candidates for jobs. Of these, 94 percent use LinkedIn to find people who might be a good fit for their organizations.[4]

One of the most powerful tools used to network online currently is LinkedIn. Since the company first launched its professional networking site in 2003, the site has grown in membership exponentially and represents a free tool that every scientist can use to build a substantial network of professional connections. LinkedIn consists of hundreds of millions of members, representing endless networking possibilities for scientists all over the world.

When you first register for the site, you will need to set up your LinkedIn profile. Be sure that your profile is complete and contains specific information about everything you have done. Include skills, tasks for every job listed, and any other experience you have had that may be relevant to potential employers.

Open your LinkedIn profile with a comprehensive summary of who you are, what you do, and where you would like to go next. Here is an example:

Inventive, passionate, and energetic scientist with industrial experience and broad expertise in diverse areas of applied physics, from optics to heat manipulation.

- International work experience in USA, UK, and Italy
- Expert on metamaterials
- Experience with prototyping and building electronic hardware for commercial and lab purposes
- Hands-on experience with micro/nano fabrications of photonics and microwave devices
- Expert in electromagnetic and multi-physics simulations
- Interested in working as a bridge between the research and industry worlds

Another critical aspect to using LinkedIn as a networking tool is group membership. At the end of every member profile, you can see which groups people belong to. Conduct a few searches on LinkedIn to find a group that matches what you are looking for, and be specific, including the type of work you would like to do and your geographic preference. For example, there is a Boston biotech group, a group for Houston energy professionals, and more. You might also join a more broad-based interest group on LinkedIn, such as PhD Careers Outside of Academia. Join as many groups as interest you, and then become an active member in a few. Try to post articles, comments, or responses when you can, as your presence in a group represents another way to build your professional identity.

My colleague Laura Stark was once moderating a panel of graduate students who had received offers for faculty positions. One of her panelists, a soon-to-be PhD, relayed how he had been invited for an interview that ultimately resulted in a job offer. He had been a frequent and thoughtful contributor to an online forum, offering many insights and helpful solutions to other scientists' problems. He had made a name for himself, and when a position became available, a fellow forum member recommended him for the position.

Q: Should I accept invitations to LinkedIn from people I don't know?

A: Check them out first, rather than accepting invitations blindly, but don't turn them down simply because you do not know them.

I am often asked whether it is smart to accept LinkedIn invitations from people you don't know. I use the following ways to gauge whether to accept a new connection. If the person uses a picture, lists descriptions of his or her work, and has more than one connection, the likelihood that the person is real seems greater. Also look for contacts that have groups, people, or interests in common with you, and seek those who might be able to benefit you, or whom you might be able to benefit. Checking out users in this way only takes a few seconds, but it helps to ensure that I am connected to real people who I might be able to help one day. I invite attendees at all of my talks to "link in" to me following my talk. If I can assist any of those who cared enough to come out and hear me speak, it would be my pleasure.

Q: When trying to connect with people on LinkedIn, should I send an email message first, or just send a LinkedIn invitation?

A: Either is fine, as long as you write a personal message *every time* you invite someone to connect via LinkedIn. Stay away from the program's default invitation. Be specific about how you know them, or why you'd like to be acquainted.

Another way to use LinkedIn is to conduct searches for organizations. If there are particular companies you are targeting, you might look for those on LinkedIn to identify current or past employees. Check to see whether you are connected in any way through other people (your first-level contacts), and request an introduction to your second- or third-level connections. I have introduced many PhDs to other professionals this way. Once you have found a company site, you can also follow the organization, read about news items, and see any current jobs posted.

Facebook, Twitter, and Blogging

Although LinkedIn is arguably the most powerful and widely used social media tool for professionals, there are other platforms that you might engage with for your job search as well.

Many users see Facebook as a primarily social networking site, but employers may browse this site as well, if they find an account associated with your name. Be sure that you are willing to have a potential employer view everything on your Facebook site—or use the appropriate privacy controls to block access. Facebook can be used quite effectively to get back in touch with family members, old friends, former classmates, and others to build relationships and your network.

On Facebook, you can use status updates or the notes field to let others know you are on the job market. I had a friend who was able to find a position successfully after moving to a new state because he posted his job search as a status update and had several positive responses and suggestions of contacts. You can also use Facebook to follow company pages where, as with LinkedIn, you can find news updates, changes in the organization, or job openings.

Twitter is another potential site for networking, offering the same opportunities—and caveats—as Facebook. Be aware of your professional identity when using this site. Try to use your real name to generate more hits when employers conduct web searches for you, and when using your real name, be professional in what you post.

Remember also that you do not need to tweet to use Twitter effectively. As with Facebook and LinkedIn, this platform is free to use, and it can be helpful in following organizations of interest. It can also be useful to follow people you have not yet connected with to stay on top of what they are doing. If you do tweet, you might consider following people who follow you. Also, join conversations around topics of interest or geographic locations, such as Clean Energy Atlanta, for example.

A proactive way for you to take control of your online identity is to start a blog. I have met several PhD scientists over the years who have done this with great success. Some have had their writing recognized and grown their network at the same time. What topics appeal to you? What might you blog about? Blogging can be a wonderful way to showcase your personality while maintaining a professional image, as blogs tend to represent a more informal—but still quite useful—platform for people of all professions and backgrounds.

Your Existing Network

Moving beyond the web, it is important for you to recognize and identify your existing, in-person network. As often as I have brought up this topic, PhDs I have worked with have claimed their personal networks to be hardly worth mentioning, bordering on nonexistent. However, if you stop to think about all of the different contacts you have made in your life, chances are you will find your *existing* personal network quite extensive (Figure 7.1).

Family and Friends

Starting with the obvious, you have your family. This includes, of course, your immediate family, but also your extended family—and that of your spouse or partner. I once worked with a postdoc who was interested in asset management. She needed a connection in the field, so I introduced her over email to my husband's uncle, who had worked on Wall Street for over twenty-five years. He was able to connect her with a few people at high-powered firms, and she ended up interviewing for a position. Think broadly, not only about your

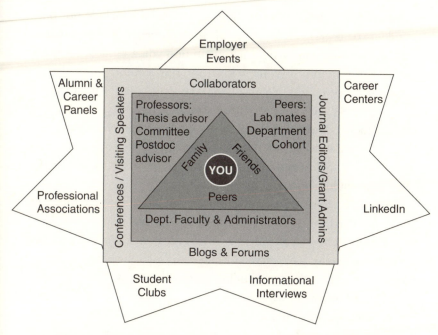

7.1. The many facets of networking

Courtesy of Laura L. Stark. Talk Your Way into a Great Job (Cambridge, MA: Office of Career Services, FAS, Harvard University, November 18, 2015), Slide 7.

nuclear and extended family, but about your partner's as well. Reach out to those in positions of interest and let them know you are seeking information, or that you are on the job market, depending on where you are in the process.

Friends are another valuable resource. Again, be sure to let your friends know what you are thinking about and where you would like to end up, and you may be surprised by the connections your friends may offer.

Education and Training

You have developed many contacts from your undergraduate institution on through your graduate school, and potentially on to your postdoc research site. Connections in this sphere include not only students or peers you knew during your program, lab, or research group, but all alumni from each institution you attended. To get in touch with alumni in your field of interest, contact your alumni association or career center. Either one may house a database of alumni who have volunteered to serve as networking contacts for all graduates. These databases are typically searchable not only by sector or career field but also geographically, so you might find alumni of your institution in Shanghai or Pittsburgh, or anywhere else you would like to end up.

In addition to students and alumni, think about the activities you were involved in during your tenure at different institutions. Can you reach out to past members or leaders of your student or postdoc groups? You may have served on the board of a postdoctoral association, for example, and it might be worth getting back in touch with board members who served at the same time, since they may be in careers or even in locations where you would like to be.

Consider reaching out to past faculty as well. It may be that you have taken courses with professors whose work you admired, and you may find that one or more have an industry collaboration that piques your interest. Send an email reminding faculty members of when you took their courses, indicate an interest in their most recent work. Let them know that you are exploring careers in industry and noticed their collaboration. Chances are that most faculty members would be willing to make an introduction for you.

Support services and offices at your institutions may also be willing and able to assist you in your search for contacts. Many career centers host career panels with current professionals and allow alumni to attend in person or online. The same may be true for postdoctoral affairs offices, so it is worth checking. Likewise, international offices may offer training in employment options for various visa types and allow current students, postdocs, and alumni to attend.

In addition to group programs, career counselors or advisors may offer individual appointments to students, postdocs, and alumni and can help you generate a list of networking contacts. These staff members often work closely with local employers and may be able to pass along the name and email address of a recruiter, scientist, or other contact at a company of interest. They may also be able to connect you with other students or postdocs going through the job search process who are focusing on the same kinds of jobs as you.

As with programs hosted by the career center, you might also think about approaching visiting scientists who give talks at your institution, whether or not you are still a student or affiliate. I once reached out to a speaker who visited my alma mater, the University of Michigan, even though I was living in the Northeast at the time. All you need is a common thread of discussion, so I opened by mentioning her visit to campus, and then confided that we were working on similar research projects and that I was struggling with a piece of mine. She responded and assisted me with my research. If you are on campus and see that a speaker will be visiting, find out whether you can join the speaker for lunch, either by contacting the hosting department or by reaching out to the speaker directly before they visit campus.

Journal Editors and Grant Administrators

If you have ever worked on a publication or grant, chances are good that you have been in touch with a journal editor or program officer. You may be curious about these jobs, in which case these contacts are directly beneficial, but even if you are not, it is important for you to remember that both of these occupations require a high degree of contact with scientists from all over the world. If you feel you have a cordial relationship with your editor or program officer, let him or her know what you are seeking, and they may be able to connect you with some relevant contacts.

Likewise, grant administrators at your research institution deal with scholars from all disciplines and may host programs on campus, bringing in speakers. They may also be helpful in pointing

USING PUBMED TO FIND A CONTACT

If you are trying to find out more about a particular organization and would like to find someone within that organization, try conducting a search on PubMed, PsychInfo, or another large database of journals in your field, using the organization name as the search term. Some PhDs have succeeded in finding the names and personal emails of those scientists who were authors on papers and work in for-profit settings, where it may be otherwise quite hard to track down a personal email address. This also provides a ready-made point of conversation, since you can mention their article in an introductory email.

you in the right direction and connecting you with professionals locally.

Professional Associations

Scientific societies and other professional associations represent a treasure trove of networking opportunities. Conferences offer many chances for you to meet with professionals in a variety of disciplines and potential career fields. Membership lists may be available to current members, and the association may offer a reduced fee to students and postdocs to join. Finally, boards of directors for professional societies are themselves a wonderful source. Board members are likely to be happy to speak with you if you express an interest in their work.

How to Approach Contacts

Having successfully identified both members of your existing network and potential future contacts, it is now time to reach out. You may experience doubt, hesitation, or trepidation. I have heard two sentiments in particular from PhDs and postdocs who are thinking about getting in touch with networking contacts: *"Why would anyone ever help me?"* and *"I don't want to bother anyone."*

Why would anyone help you? Well, have you ever helped anyone? Have you answered questions at your poster during a conference? Agreed to meet with an undergraduate considering your graduate program? Spoken with graduate students you didn't know who were interested in applying for a postdoc at your institution?

Indeed, you may have even hosted a prospective graduate student you had never met during an admissions event or weekend. And why do you think you were willing to engage in these activities? Perhaps because the people who were reaching out to you regarded you as an expert in the topic at hand: your own poster, your undergraduate program, or your graduate institution.

Similarly, reaching out to a professional through networking suggests that you consider that person to be an expert in his or her own work, and many, if not most, professionals will find this expectation flattering. Contacting professionals directly makes sense because we know that the best source of information about a particular occupation or career field is most often someone who is in that field currently.

Contacting a professional suggests you consider that person an expert in his or her field.

Even if you do consider current professionals to be experts in their knowledge of their own job, you may still hesitate to contact them, believing that they have more important work to do than to talk with you about their jobs. It is likely that they received similar aid early in their careers, and many people are interested in returning the favor in the same way, whether they know you or not. Indeed, postdocs and graduate students I have worked with have often been overwhelmed by the generosity shown by professionals they have never met.

When I was working on this book, I contacted several professionals involved in the *Mathematics in Industry* study by the Society for Industrial and Applied Mathematics.[5] Not only were they willing to help me as I navigated their research project, they were quick to suggest a phone call or Skype meeting rather than email, so that we could meet one another face to face and cover more ground. In spite of my time spent teaching students and postdocs the value of reaching out to strangers, I still found their generosity astonishing—and I am so grateful for it.

It is time now to determine how best to connect with someone you know very well, have met on occasion, or don't know at all. Begin any conversation by expressing an interest in the other person. This engages your contact immediately and will most likely keep the conversation going. Here are more specific tips for reaching out to anyone in your network:

1. *Email first.* Most people prefer to receive an email to a phone call, since this mode of communication offers people a chance to respond at a time that works best for them. In your email, include a brief introduction of yourself, what you find interesting about their work, and what kind of information you are seeking. Request to meet for coffee or, even better, at their workplace, or suggest a thirty-minute telephone call, if that is easiest for the contact. Your email might look something like this:

   "Hello, Dr. Z, My name is Marika Eisenstein, and I am currently a postdoctoral fellow at the University of Washington, conducting research in cellular neurobiology. I am considering the field of technology transfer and found your background relevant to my interests. I have a few questions about the field and wondered whether you might be willing to set up a brief phone call to discuss your work? I am free most afternoons and am happy to contact you at a time and date that work for you. Thank you in advance for your time!"

2. *Follow up* with another email or phone call in two to four weeks, if they don't respond. Often times, your email is not a first priority, but people really do want to help. By embracing the "polite persistence" approach, you can get through to even the busiest people.

3. *Keep track* of your contacts, how you met them, who referred you to them, what you spoke of, who they suggested you

contact, and so on. Use a spreadsheet or other system to track these conversations.

Once you have scheduled a time to talk with networking contacts, be clear about your goals and have questions ready. It often helps to begin by asking about a person's own story, education, background, and how they entered this particular field. As you listen to these stories, you might identify gaps in your own training or experience that have been a common thread in the stories of multiple professionals. Understanding where your gaps are and what additional training you might need will help you to strengthen your candidacy for particular jobs.

Anytime you speak with a contact, be sure to send a thank-you letter or email as soon as possible after the meeting. It is critical to follow this protocol to engender positive feelings about the meeting, to be gracious and courteous, and to remain in your networking contact's consciousness.

Differences between Networking and Informational Interviewing

Both networking and informational interviewing are critically important for your job search, and they share much in common. In both cases, you are reaching out to contacts, either for information or potentially for connecting with other people. The crucial difference between these two is *timing*. I strongly encourage you to engage in informational interviewing *before* you actively look for a job. I suggest this because it tends to feel less awkward and will induce less stress if you are truly in an information-gathering stage. Networking itself is a more direct approach, and carries with it a greater sense of urgency. It is still aimed at gathering information, but with more immediacy.

The language you use when contacting people will vary depending on your own timing. Do you have twelve months to identify options, make a decision, and find a job, or do you have only twelve weeks? If you have more time to explore and investigate options, let the professional you are speaking with know that—and be sure to stress that you are gathering information and *not* looking for a job.

If you have only a few weeks before you defend your thesis, or before your postdoctoral appointment ends, your language should change slightly. While you can share that you are on the job market, stress that you are not expecting the person to provide you with a job. Suggest that you are gathering information about their work, looking for additional contacts, and hoping to make use of any leads they may have or insights they can share about trends in the field. If you have cultivated a network over time through informational interviews, you can reactivate it when you are on the market. This approach will feel more comfortable than approaching people for the first time when you are under pressure to find a job quickly.

Still, many of us may find ourselves in that very position, knowing we have a limited time period within which to find work, and that is okay—as long as you don't open networking conversations by asking whether an organization is hiring. If the answer to this question is "no," it is very difficult to recover the conversation. However, if you use the approach above, expressing an interest in the other person and his or her work, that opens the conversation and encourages a dialogue that can continue organically. For example, try beginning with, "I see that you went to the University of Alabama for your PhD. Did you work with Professor Z?"

Never open networking conversations with "Are you hiring?"

Always remember to end every networking conversation by asking whether there is another professional you should contact. In this way, you can create a domino effect by being referred to a new connection every time you speak with someone.

For more ideas and conversation starters, you can refer to the list of informational interviewing questions in Chapter 6. Some of these will work for networking conversations and some may not, but the prompts listed will at least get the conversation going.

Mutual Benefits of Networking

Some PhDs may view the networking process as one-sided, where they take what they need from a contact and give nothing in return. That view, while understandable, is nevertheless untrue. Both parties engaged in a networking interaction tend to benefit, if not directly or immediately, then indirectly or eventually. Here are few illustrations of these benefits.[6]

Benefits to You
- Gain firsthand knowledge of a work environment, enabling you to determine whether it is a fit for you personally—or not
- Collect insider advice on how to enter a field, and potentially advice on a resume or CV
- Learn the vocabulary, vernacular, and culture of the field
- Understand current trends
- Generate additional career options
- Build networks through further contacts
- If you maintain relationships, your contacts may alert you to job opportunities.

Benefits to Your Networking Contact
- Talk about your self and your own experience
- Share wisdom
- Offer the same assistance received early in your contact's career
- Meet a new colleague
- Check out a prospective employee without commitment, test the chemistry and fit for the organization
- Save time, money, and effort in finding a suitable employee

Although these benefits occur naturally over the course of a conversation or relationship, you should also make every effort to provide direct benefits to your networking contacts as often as you can. For example, you might share an article or resource related to a past conversation, put that person in touch with another PhD who may have similar career interests, or help them with something unrelated to work. I once mentioned to a recruiter that I like to hike with my young boys, and he sent me an email a few weeks later, listing hikes near me that were a good fit for families.

Each time you endeavor to help someone in your network, you bring yourself back into the other person's consciousness, thereby increasing the likelihood they will think of you when a relevant opportunity arises in the future.

Maintaining Relationships over Time

Keep up relationships with people in your field of interest. While this might sound like a daunting task, you can limit the number of people you get in touch with and the number of times you contact them. Choose people you feel strongest about keeping in touch with, and try to reach out to them at various milestones: for example, you might email them if you have a paper accepted, if you just defended, if you presented a poster, or if you are traveling to a meeting in their area. Be strategic about your follow-up emails and contacts, and don't think you need to keep in touch with everyone in your network. But do make a point to keep in touch with those who can help you with your long-term goals. Keep them up to date on your work.

Remember: *You're either networking, or you're not working.*

PART THREE

Getting Started on Your Job Search

CHAPTER 8

How to Craft Your Individual Development Plan

In Part One, we took a close look at your personal profile, including your interests, skills, and values as they pertain to careers. In Part Two, we learned about occupational requirements and the different sectors that employ science PhDs, and how to build skills, experiences, and a network to move into those fields. In this first chapter of Part Three, we will determine how to synthesize this information and, once we do, how to make a career decision and set goals for your job search.

There are several tools we can use to bring this information together in a way that will assist us in making a decision. To start, we can learn about a crucial tool, the Individual Development Plan (IDP). Next, we can reflect on your work in previous chapters to create a career development profile. Then we can use an occupational assessment chart to decide which career fields might represent a good fit for you. Finally, we can draft an IDP that incorporates not only your career goals, but also your research goals and plans for the remainder of your program or postdoctoral appointment.

Your Individual Development Plan

The Individual Development Plan (IDP) emerged from a meeting of the Training and Careers Committee of the Federation of American

Societies for Experimental Biology (FASEB) in 2001. They were looking for tools that would help science trainees move forward not only with their research, but with their career plans. The IDP was designed to assist graduate students, postdoctoral scholars, and their advisors and PIs to identify developmental needs, to enhance communication, and to draft shared goals.

Julian Preston, an early adopter of the plan, describes the IDP this way:

> Each of the three component words in the IDP provides a measure of the value of the plan: *Individual* emphasizes the need to consider the unique training and career goals of each fellow. *Development* stresses the identification of steps needed to achieve the goals. *Plan* stresses the specific steps needed to reach career goals rather than relying on the more traditional random walk.[1]

Ideally, the trainee and PI would collaborate in crafting a plan.[2] While this type of exchange may not occur, it is crucial for all trainees to reach out and try to communicate expectations to a faculty advisor, both for their research and for their career development. If it is easier to begin with a conversation about research goals, by all means, that is a great place to start. However, if you have a solid understanding of your career options and know that you will need to develop skills and engage in experiences to make yourself more marketable, a conversation between you and your advisor on that topic is probably inevitable.

Steve Linn, a former postdoc who has used an IDP, explains, "One of the most useful aspects of the IDP is that it provides an opportunity for the expectations of both postdoctoral fellows [or graduate students] and their advisors to be explicitly stated and for a structural framework to be laid out for achieving the desired goals," he says. "I've observed in the past that those things are too often left to casual verbal exchanges, which can produce a lot of misunderstanding and leave both parties frustrated."[3]

Using an IDP, then, should free up channels of communication between the PhD or postdoc and the mentor. Let's now take a closer look at its various sections.

Components of an IDP. According to the outline created by FASEB, a series of steps should be taken by both the trainee and the advisor in the creation and follow through of an individual development plan, as shown in Figure 8.1. FASEB went beyond the outline in this figure and suggested tasks for each stage, including an evaluation of skills, interests, values, and development areas for the first step, identifying career opportunities as well as skills gaps for the second step, and sketching a timeline for the third step that would include specific goals and deadlines for developing the skills and experiences needed to achieve long-term research and career goals.[4]

Creation of myIDP. Fast forward ten years, and authors Philip Clifford, Bill Lindstaedt, Cynthia Fuhrmann, and Jennifer Hobin

	For Postdoctoral Fellows and Graduate Students	For Mentors
Step 1	Conduct a self-assessment	Become familiar with available opportunities
Step 2	Survey opportunities with mentor	Discuss opportunities
Step 3	Write an IDP Share IDP with mentor	Review IDP and help revise
Step 4	Implement the plan Revise the IDP as needed	Establish regular review of progress Help revise the IDP as needed

8.1. Four steps to a career plan for graduate students, postdocs, and mentors (faculty advisors or PIs)

Adapted from the Federation of American Societies for Experimental Biology (FASEB).

transferred this process to an online platform, entitled *myIDP*. An exceptional resource, it is available at no charge to scientists all over the world, and is approaching 100,000 users. This interactive, web-based career-planning tool can help you engage in the self-assessment process, explore career options, set goals, and put your plan in place. It can be used to learn more about each of these stages, as it includes several links to insightful articles, but can also be valuable in keeping trainees on track, as it allows users to define goals and attach deadlines to these goals. The section of myIDP that I use most often is the section under Career Exploration entitled "Read about Careers." This section includes a tab called "Resources" that lists twenty different career fields and links to articles, websites, and professional organizations for all of these fields.

Requirement of IDPs by federal agencies. In 2009, the National Science Foundation (NSF) began to require a mentoring and professional development plan for all postdoctoral researchers funded by the NSF. Since that time, research institutions have had to compile a document outlining the specifics of how mentorship will be handled for funded scholars and how their professional development needs will be met. And, in 2014, the National Institutes of Health (NIH) mandated the use of IDPs at all research institutions funded by their institutes. According to the NIH, "annual progress reports received on/after October 1, 2014 must include a section to describe how Individual Development Plans (IDPs) are used to identify and promote the career goals of graduate students and post-doctoral researchers associated with the award."[5]

Institutional use of the IDP. Since the NIH issued its requirement, several institutions have incorporated the use of IDPs into training programs. For an example of how one university approaches this exercise, visit the website of the Office of Biomedical Research Education and Training at Vanderbilt University's School of Medicine. This office requires not only the completion of an IDP for each incoming postdoc but an updated version, with mentor comments, every year. In fact, they will not issue a reappointment letter without it.[6] This office should serve as a model to other universities and research institutes considering how to fold the IDP into institutional policies.

Now that we have explored the history and use of the IDP in the scientific community, let's take a closer look at how you might use it going forward.

Exercise: Choosing Careers

Review the responses to the self-assessment exercises you completed in Chapters 1–3, your career development profile in Chapter 3, careers of interest identified in Chapter 5, and occupations brainstormed with a friend in Chapter 6. Now you can determine how well your careers of interest fit your personality, and how willing you are to work to develop skills, networking contacts, or other steps needed to move into these particular occupations.[7]

Using three occupations identified by you, a friend or colleague, or through your exploration in Chapter 6, copy the grid in Figure 8.2 into your career notebook and complete the top of this chart. Next, answer each question to the left of the grid for each occupation listed by putting a checkmark next to Yes, No, or Unsure, referring to your career development profile as needed. Finally, tally the number of checkmarks for each occupation to give yourself a better sense of the fit of certain occupations, as well as areas needing development.

Take a look at your scores. Which occupations seem a better fit for you at this point? What skills do you need to develop? Do you need to build a more robust network in your field of interest? Now that you have a better sense of which occupations might be a potential fit for you and what you may need to accomplish to enter these careers, it is time to develop a plan of action.

Drafting an IDP Using SMART Goals

If you have completed the exercises in previous chapters and the decision-making exercise in this chapter, you have finished steps one

Is this occupation a fit for:	Occupation 1			Occupation 2			Occupation 3		
	Yes	No	Unsure	Yes	No	Unsure	Yes	No	Unsure
My interests? Would I like this work well enough to make it my career?									
My current skills?									
Do I have the potential to develop them?									
My values? Work setting or working conditions that I'd prefer?									
My current networking contacts?									
Do I have the potential to develop them?									
My geographic preferences?									
My visa status?									
My personal and financial needs?									
TOTALS:									

8.2. Occupational assessment chart

and two of the IDP process. You most likely have emerged with some goals in mind—or at least identified gaps to be filled. What are these gaps for you? Do you need to develop technical skills? Might you need to conduct more informational interviews? Do you still need to make a decision between a few career options? Or do you know precisely what you would like to do but need a roadmap to get there? In any of these cases, you can use an IDP to sketch out the specifics.

In the exercise that follows, you will be asked to choose a long-term career goal, assign a deadline to your long-term goal, and come up with several related short-term goals and tasks. You can—and should—use the same process for setting your research goals, and these should be incorporated into your final IDP. The exercise that follows simply lays out a strategy for setting goals, using a career-focused example.

You can engage in this goal-setting exercise with your mentor or on your own, but be sure to review your finished plan together. It is crucial for you to keep the lines of communication open, and that means sharing your research and career goals with your faculty advisor. If you do not feel comfortable sharing your IDP with your current PI, share it with another mentor or a friend. The more readily you can talk about your plan and the steps needed to accomplish it, the more likely you are to move forward with it, as more people will be aware of your goals and will be able to keep you accountable to them.

As you begin to think through your goals and associated tasks needed to complete those goals, it is important to ask yourself: Are they SMART? *SMART* stands for:

Specific—Do I know precisely what has to happen?
Measurable—Will I know whether I've completed this task?
Achievable—Is it realistic or do-able?
Result-oriented—Will it really move me toward my goal?
Time-limited—Does it have a due date?

If the tasks related to your long-term goals are SMART, it will be much easier for you to work through them and to know when they have been achieved.

Exercise: Goal-Setting for your IDP

Now you are ready to create your IDP. First, choose a long-term goal and write it down. Set a deadline for this goal, then break your long-term goal into short-term goals. Define tasks with deadlines for each short-term goal. Here is a partial IDP, listing only a long-term career goal. A full IDP would include a long-term research goal as well.

Long-Term Goal
To find a faculty position at a liberal arts college
Deadline: August 2017

Short-Term Goals
  A. Network with faculty at primarily undergraduate institutions.
  B. Strengthen teaching skills and experience.
  C. Build teaching portfolio.
  D. Find funding for undergraduate-friendly research.

Tasks and Deadlines
  A.1. Email three faculty at local colleges and suggest meeting for coffee on their campuses (June 15).
  B.1. Contact chair of science department at local college and volunteer to serve as guest lecturer (August 15).
  C.1. Attend course on assembling a teaching portfolio at a Center for Teaching and Learning (Sept. 9).
  D.1. Create automated email alerts for grants funding undergraduate research using the Pivot grants database (Sept. 30).

As you can see from the sample, each short-term goal has a corresponding task with an attendant deadline. In reality, each of your short-term goals will require several tasks to complete—this sample is merely an illustration.

Breaking down your research and career goals will prove more gratifying than setting broad goals that are neither reasonable, measurable, nor time-delimited. For example, saying "I will work on my job search this weekend" is a recipe for feeling frustrated and defeated by the end of the weekend. You may feel overwhelmed, not knowing where to begin, and you certainly won't know by the end of the weekend whether you have accomplished your goal because your sweeping statement leaves nothing concrete to measure.

If instead you break down your long-range goals into short-term, manageable chunks, you will feel more in control and less anxious, and your progress will be more apparent. With this in mind, answer the following questions in your career notebook or on your laptop:

1. What is your long-term career goal? (This might be a specific career, or your goal may be simply to *find* a long-term career goal.)
2. What is your deadline for this goal?
3. What steps do you have to take in order to accomplish this goal? (What are your short-term goals?)
4. List at least three specific tasks associated with each short-term goal, and attach a deadline to each task.

Once you have drafted your IDP, sketch out your goals on a calendar, date book, smart phone, or computer. There are also many applications available that can keep you accountable to the goals and tasks you set, including myIDP. Try using an application if you think it would help you to stay on track.

Sharing Your IDP with Your Advisor

If you did not map out your IDP together with a mentor and are now thinking about sharing your plan with your primary faculty advisor, there are several things you can do to prepare for this meeting.

1. Share your written IDP with your advisor in advance of the meeting, stressing that this document is a work in progress and that you are seeking feedback on it.
2. Consider what you would like for the outcome of the meeting to be.
3. Before the meeting, think carefully not only about your goals, but about the goals of your research group or lab, and of your advisor. Try to anticipate any concerns your advisor might have about your research and career development plans.
4. With the goals of the research group and your advisor in mind, start the meeting by explaining how your research and career goals fit into the overall plan for the lab or group.
5. Stress your commitment to completing any work you and your advisor had already planned, and then move into a discussion about new and updated projects and research goals outlined in your IDP.
6. Be sure to engage in active listening, paying attention to any issues or concerns your advisor wants to address. Be flexible when discussing possible compromises or changes to your plan.
7. If you feel comfortable, move into a discussion about your long-range career goal, and what steps you might need to take (such as additional training or an internship) to move into that role. Again, listen closely to your advisor's reaction to this news. Do not be dismayed if the response is not immediately positive, since the thought of your departure may be difficult to process right away. Your advisor may need time to digest your plans.

Some people may feel at ease discussing their career goals with their PIs and some may not. If you are not comfortable having this conversation with your primary advisor, have it with another mentor. It is critical to your success that you surround yourself with mentors who can and will support you throughout the career development process. These mentors may certainly offer advice, but they can also keep you on track if you meet with them regularly to discuss your plans.

Reviewing and Revising Your IDP

Step 4 of the IDP outline in Figure 8.1 suggests that a regular review of your progress take place, and that, based on this review, an IDP be revised and updated to reflect the current state of your goals and plans. This type of review may occur every six months or yearly, as long as it does indeed take place. It is your responsibility to make this meeting happen. You are the person who has the ultimate responsibility for your own career, not your advisor. Be sure to schedule this meeting at regular intervals to ensure that you stay on track with your research and career goals and keep the lines of communication open between yourself and your PI.

When scheduling this meeting, you should put aside time to review and update your existing IDP in advance. If you find after six months to a year that you have made progress and met some short-term goals, good for you! Take a moment to reflect on what made you successful, celebrate, and then move on to new tasks.

Barriers to Career Development

If you find that you are not making progress, it will be important for you to identify potential barriers to your work. The Center for Professional Development at the University of Hartford developed the following exercise on identifying and overcoming any obstacles you might face to meeting your goals.

Barriers are those things that block your way to achieving your goals or proceeding with your plans. They stop you from taking action about your future. These barriers can be classified into the

Type	Your Barriers	Your Strategies
Economic		
Psychological		
Physical		
Educational		
Social		

8.3. Barriers to career development

categories of economic, psychological, physical, educational, and social. Economic barriers are financial restrictions on your actions. An example would be your desire to take a new job but your inability to afford a salary lower than what you are currently paid. Psychological barriers are personal issues that restrict you from taking action. Some examples might include procrastination, fear of failure, difficulty making decisions, or feeling overwhelmed. Physical barriers are things that you think may limit you, such as age, gender, or chronic physical challenges. Educational barriers may prevent you from pursuing a position because you think you lack an appropriate degree or have outdated skills. And finally, social barriers exist in your chosen obligations to others. You may have to fulfill your role as a parent, a spouse, or a partner, which requires attention, presence, and time. You may be a member of a group that has certain rules that restrict your choices. Addressing these barriers and finding solutions or ways of coping with them will make your path to effective career development easier.

Exercise: Overcoming Barriers to Career Development

Think of barriers in each of the categories that apply to your situation. Copy the grid in Figure 8.3 into your career notebook or laptop, and write your specific barriers under each category in the left-hand column. After you have noted your barriers, identify possible solutions or strategies for addressing each one. Write your strategies in the right-hand column.

Addressing any barriers to your progress either in your research or for your long-term career goals will allow you to move forward and continue on a path to securing the job you desire. Stay focused and continue to review and revise your plan as needed, with the help of your faculty advisor, peers, career counselors, or other mentors to ensure your success.

CHAPTER 9

How to Apply for Jobs

Whether you feel confident about the career you would like to pursue or are still exploring your options, you will need to develop a profile of your work on paper and online that can be shared with others. The primary documents used in job searches are resumes, CVs, and cover letters. In this chapter, we will take a closer look at these documents and then move on to other documents required for more specific searches.

CVs and Resumes

"What is the difference between a CV and a resume, anyway?" This question arises quite frequently among graduate students and postdocs, but the answer is fairly straightforward: these two documents differ in purpose, content, and length. The purpose of a CV, or Curriculum Vitae, is to illustrate your entire academic and work history. As such, there is no limit to the length of this document. CVs are most often used for faculty positions, sometimes for research positions, and are often required for funding applications. Resumes, on the other hand, are used to apply for nonfaculty jobs and are limited in length, usually to two pages for graduate students and postdocs, although they could possibly be a bit longer.

Variations in the content of these two documents tend to vex researchers who are going on the job market for the first time, but there are some general guidelines I can offer that will assist you in developing content and format for either.

Drafting a CV

When drafting your CV, begin by listing all of your degrees, research appointments, teaching experience, past jobs, awards, volunteer activities, grants, professional societies, and more. Do not try to edit your experiences at this point—it is important to create a master version of your CV that covers your entire background, and you can always edit this document later to fit a particular job you are applying for. Once you've listed as much as you can, try to group together like entries among your experiences using the category headings below to describe these groups:

1. Name, address, city, state, zip code, phone, email, and website (if applicable)
2. Education, dissertation, master's project, thesis, professional competencies
3. Areas of expertise, areas of concentration in graduate study, internships
4. Teaching interests, teaching experience, courses taught
5. Research interests, research assistantships, postdoctoral experience, research appointments, research experience, professional experience
6. Academic appointments, professional summary, related experience, administrative experience, consulting experience
7. Academic service, advising, university involvement, outreach, leadership, university assignments, professional development
8. Professional association advisory boards, advisory committees, national boards, professional activities, journal reviews, conference participation, conference presentations, conference

leadership, workshop presentations, invited lectures, lectures and colloquia

9. Publications, abstracts, scholarly works, books, chapters, editorial boards, professional papers, technical papers, refereed journal articles, editorial appointments, articles and monographs, book reviews
10. Research grants, funded projects, grants and contracts, patents
11. Awards, scholarships, fellowships, honors, activities and distinctions
12. Professional recognition, prizes, professional memberships, affiliations, memberships in scholarly societies, professional societies and organizations, and honorary societies
13. Professional certification, clinical certification, licensure, endorsements, special training
14. Foreign study, study abroad, travel abroad, international projects, languages, language competencies

Now that you have listed all of your experiences and chosen logical category headings, it is important to format your CV so it is accessible and readable. To begin, consider the following formatting tips. Begin this document with your name and contact information, *not* the words "Curriculum Vitae," and set up the page with margins between one and one and a half inches, leaving sufficient white space between categories and category entries. A header or footer with your name and page number on every page but the first will ensure that pages do not become lost or out of order. Use boldface for the most important information, such as job title, and use bullet points rather than paragraphs to describe what you've done, listing dates on the right-hand side.

As you draft your CV, avoid the first person ("I," "my"). Include for each job or position held your title, the institution, city, state, country (if not the United States), and inclusive dates. Do not forget to include your thesis title and advisor's name.

In describing what you have done in each position, use short phrases beginning with strong action verbs. Job descriptions on CVs should not be lengthy—indeed some hiring committees for faculty positions may wish only to see your title, where you conducted your research, and your advisor's name. Still, for most readers, it will be helpful for you to include a bullet point or two describing your research in your graduate program and in your postdoctoral work, if applicable.

CVs can be tailored to particular positions of interest as well. When reviewing academic job openings, read the job ads carefully, highlighting areas of importance to the committee. You can use this information to craft your brief bullet-point phrases on your CV to emphasize your suitability for the position.

If the faculty job description is sparse or nonexistent (some universities publish only the level of the post and the department), you will have to do some background research in order to strengthen your CV and other parts of your application. Study the website of the department in question, and in particular, the research areas of current faculty. It will be incumbent upon you to demonstrate that your research is not only compelling, but that it will meet a need currently unmet in the department.

When compiling your experiences on your CV, think carefully also about the order in which categories appear. If, for example, you are applying for a faculty position at a teaching-focused four-year or community college or a teaching-centered fellowship, you should stress teaching first and foremost on your CV. In this scenario, you'll want to order the categories to promote your teaching experiences and interests on the first page:

Education
Teaching Interests
Teaching Experience
Research Experience

. . . and so on.

Conversely, if you are applying for a research position in a government institution or in industry, for a faculty job at a large research-driven institution, or applying for a grant or fellowship focused on your research, you will need to highlight your research on the first page of your CV. You may even want to move your publications forward (typically this category falls toward the end of the document). Some PhDs applying for faculty jobs might create two separate documents—one for teaching positions and one for research positions—and use the one most appropriate for what they are applying for.

When preparing to apply for faculty jobs, you should come to learn the most generally accepted practices for CV writing in your field. Look for CVs of junior faculty who were recently hired at the types of institutions you would consider. Reviewing the CVs of recent PhDs will help you assemble your own document and assess your competition. Take a look at the CV for a typical faculty applicant shown on the following pages, for example. Who is being hired, and why? What makes their CVs attractive? Are they investigating an especially captivating area of research? Do they have numerous high-impact journal articles? Are they coming in with years of their own funding? Being armed with this information will help you to make decisions about your career goals and fill gaps in your own candidacy, enabling you to be the most attractive candidate possible in this tight job market.

Once you have completed a draft of your CV, be sure to share it with *many* people: your advisor, colleagues, friends, partners, other faculty members, career counselors, advisors, and more. It is crucial that this document be easy to read and error-free. It needs to make sense to any reader, so sharing it with as many people as possible will help you put your best foot forward in any application process.

Converting a CV to a Resume

Now that you have a version of your CV to use for funding, research jobs, or faculty applications, it may be worthwhile to create a resume to use for jobs outside of faculty and teaching appointments. To build a resume, it is helpful to use your CV as a baseline

SAMPLE CV OF A FACULTY JOB CANDIDATE

Benjamin E. Wolfe
Harvard University
FAS Center for Systems Biology
Cambridge, MA
benjamin.wolfe@fas.harvard.edu

Linking ecological patterns and processes with the genes and genomes of microbial communities.

Passionate promoter of microbial literacy through teaching and writing.

EDUCATION and TRAINING

Postdoctoral Researcher, FAS Center for Systems Biology, **Harvard University,** 2011–2014
Ph.D., Organismic and Evolutionary Biology, **Harvard University,** 2005–2010
M.Sc., Department of Botany, **University of Guelph,** 2003–2005
B.S., Natural Resources/Plant Science, *magna cum laude,* **Cornell University,** 1999–2003, *Honors Thesis Advisor*: Dr. Barbara L. Bedford

PUBLICATIONS

Manuscripts published in peer-reviewed journals
Wolfe, BE, JE Button, M Santarelli, RJ Dutton. 2014. Cheese rind communities provide tractable systems for *in situ* and *in vitro* studies of microbial diversity. *Cell* 158: 422–433.

Hess, J, I Skrede, BE Wolfe, K LaButti, RA Ohm, IV Grigoriev, and A Pringle. 2014. Transposable element dynamics among asymbiotic and ectomycorrhizal *Amanita* fungi. *Genome Biology and Evolution* 6: 1564–1578.

Lawrence, DA, CF Maurice, RN Carmody, DB Gootenberg, JE Button, BE Wolfe, AV Ling, S Devlin, M Fischbach, SB Biddinger, RJ Dutton, PJ Turnbaugh. Diet rapidly and reproducibly alters the human gut microbiome. *Nature* 505: 559–563.

Cheng-Chih H, MS ElNaggar, Y Peng, J Fang, LM Sanchez, SJ Mascuch, KA Møller, EK Alazzeh, J Pikula, RA Quinn, Y Zeng, BE Wolfe, RJ Dutton, L Gerwick, L Zhang, X Liu, M Mansson, and Pieter C. Dorrestein. 2013. Real-time metabolomics on living microorganisms using ambient electrospray ionization flow-probe. *Analytical Chemistry* 85: 7014–7018.

Wolfe, BE, RE Tulloss, and A Pringle. 2012. The irreversible loss of a decomposition pathway marks the single origin of an ectomycorrhizal symbiosis. *PLoS One* 7(7): e39597.

Wolfe, BE, M. Kuo, and A Pringle. 2012. *Amanita thiersii* is saprotrophic and expanding its range in the United States. *Mycologia* 104: 22–33.

Wolfe, BE, and A Pringle. 2012. Geographically structured host specificity is caused by the range expansions and host shifts of a symbiotic fungus. *The ISME Journal* 6: 745–755.

Wolfe, BE, F Richard, HB Cross, and A Pringle. 2010. Distribution and abundance of the introduced ectomycorrhizal fungus *Amanita phalloides* in North America. *New Phytologist* 185: 803–816.

Vellinga, EC, BE Wolfe, and A Pringle. 2009. Global patterns of ectomycorrhizal introductions. *New Phytologist* 181: 960–973.

Rodgers, VL, BE Wolfe, L Werden, and AC Finzi. 2008. The invasive species *Alliaria petiolata* (garlic mustard) increases soil nutrient availability in northern hardwood–conifer forests. *Oecologia* 157: 459–471.

Peterson, CN, S Day, BE Wolfe, AM Ellison, R Kolter, and A Pringle. 2008. A keystone predator controls bacterial diversity in the pitcher plant (*Sarracenia purpurea*) microecosystem. *Environmental Microbiology* 10: 2257–2266.

Wolfe, BE, VL Rodgers, KA Stinson and A Pringle. 2008. The invasive plant *Alliaria petiolata* (garlic mustard) inhibits ectomycorrhizal fungi in its introduced range. *Journal of Ecology* 96: 777–783.

Wolfe, BE, MC Rillig, DL Mummey, & JN Klironomos. 2007. Small-scale spatial heterogeneity of arbuscular mycorrhizal fungi in a calcareous fen. *Mycorrhiza* 17: 175–183.

Wolfe, BE, PA Weishampel, & JN Klironomos. 2006. Arbuscular mycorrhizal fungi and water table affect wetland plant community composition. *Journal of Ecology* 94: 905–914.

Stinson, KA, SA Campbell, JR Powell, BE Wolfe, RM Callaway, GC Thelen, SG Hallett, D Prati, and JN Klironomos. 2006. Invasive plant suppresses the growth of native tree seedlings by disrupting belowground mutualisms. *PLoS Biology* 4: 727–731.

Wolfe, BE, and JN Klironomos. 2005. Breaking new ground: soil communities and exotic plant invasion. *BioScience* 55: 477–487.

Wolfe, BE, BC Husband, and JN Klironomos. 2005. Effects of a belowground mutualism on an aboveground mutualism. *Ecology Letters* 8: 218–223.

Klironomos JN, MF Allen, MC Rillig, J Piotrowski, S Makvandi-Nejad, BE Wolfe, and JR Powell. 2005. Abrupt rise in atmospheric CO_2 overestimates community response in a model plant-soil system. *Nature* 433: 621–624.

Scholarly book chapters

Wolfe BE, JL Parrent, AM Koch, BA Sikes, M Gardes, and JN Klironomos. 2009. Spatial heterogeneity in mycorrhizal populations and communities: scales and mechanisms. In *Mycorrhizas—Functional Processes and Ecological Impact*. Edited by C. Azcon-Aguilar, J.M. Barea, S. Gianinazzi, V. Gianinazzi-Pearson. Springer-Verlag. Berlin, Heidelberg. Chapter 12. Pg. 167–218.

RESEARCH GRANTS

National Science Foundation Doctoral Dissertation Improvement Grant. $12,000. The genus *Amanita* as a model for the evolution of symbiosis. 2008–2010.

Deland Award, Arnold Arboretum, Harvard University. $5,000. Ecology and evolution of the ectomycorrhizal symbiosis in *Amanita*. 2007.

New England Botanical Club, Graduate Student Research Award. $2,000. Biogeography, genetic diversity and host specificity of *Amanita* in New England. 2007.

Mycological Society of America, Clark T. Rogerson Research Award. $1,000. Origins and diversification of symbiosis in *Amanita*. 2007.

Mycological Society of America, Forest Fungal Ecology Research Award. $1,000. Arbuscular mycorrhizal fungi of alvar ecosystems. 2004.

Cornell University, Program in Biogeochemistry and Environmental Change Research Grant. $3,000. Effects of arbuscular mycorrhizal fungi and phosphorus enrichment on growth and nutrient uptake of *Solidago patula*. 2002.

TEACHING and MENTORING EXPERIENCE

Instructor, *Feast and Famine: The Microbiology of Food*, Harvard Summer School, 2012–2013

Director, Harvard *Microbial Sciences Initiative Undergraduate Fellowship Program*, 2012–2013

Teaching Fellow, *Debunking Biology Myths*, Harvard University Extension School, 2012

Teaching Fellow, *Evolutionary Biology*, Harvard University, 2010
Teaching Fellow, *Biology of the Fungi*, Harvard University, 2006, 2009
Teaching Assistant, *Population Ecology*, University of Guelph, 2004
Teaching Assistant, *Lab & Field Methods in Ecology*, University of Guelph, 2003

Supervised the following undergraduate researchers as a postdoc or graduate student at Harvard University:
- Shanice Webster, Grinnell College (visiting student in summer program), 2013
- Juan Alvarez, University of Maryland (visiting student in summer program), 2014
- Adriann Negreros, Harvard College, 2011
- Kristi Fenstermacher, Harvard College, 2007

AWARDS and FELLOWSHIPS

Bowdoin Prize for Graduate Essays in the Natural Sciences, Harvard University, 2010
Derek Bok Certificate of Distinction in Teaching, Harvard University, 2007
Runner-up for **John L. Harper Young Investigator's Prize,** *Journal of Ecology*, 2006
Soil Ecology Society Meeting—Student Oral Presentation Award (3rd Place), 2005
National Science Foundation Graduate Research Fellowship, 2004–2009
Richard Church Senior Service Award, Cornell University, 2003
Howard Hughes Medical Institute Undergraduate Scholar, Cornell University, 2002
Morris K. Udall Foundation Scholarship in National Environmental Policy, 2001–2002

SCIENCE OUTREACH

Writing

MicrobiaFoods.org—digesting the science of fermented foods
Co-founded, designed, and write for this website with Bronwen Percival, a cheese buyer for Neal's Yard Dairy in the United Kingdom. The website provides free accessible summaries of scientific literature on the microbiology of fermented foods. 2014.

"We Smell Sea Smells" *Lucky Peach* magazine. Issue 12. 2014
"Tales from the Yeast." *World of Fine Wine*. Issue 41. 2013
"Chefology"—An online series on the biology of food for *Boston*
    magazine
"American microbial terroir." *Lucky Peach* magazine. Issue 4. 2012

Lectures and workshops

Scientific advisor for Long Island Mycological Club, Long Island, NY
    — Scientific advisor for amateur mycological club, providing DNA
       identification of field collected-specimens and contributor to
       newsletter, 2008–present

Formaggio Kitchen, Cambridge, MA
    — Gave lectures on the microbiology of food at classes for food
       enthusiasts, 2013

Scientific advisor for Momofuku Culinary Lab, New York City, NY
    — Scientific advisor for projects involving the microbial ecology of
       food and author of articles for *Lucky Peach* magazine about food
       microbiology, 2010–2012

**MIT Department of Biology Summer Workshop for High School
    Teachers, Cambridge, MA**
    — Guest lecturer on cheese microbial communities, 2011

Harvard Yard Soils Project, Harvard University, Cambridge, MA
    — Conducted surveys of the microbial diversity of Harvard Yard
       to assist with the transition to organic landscaping management,
       2010

Microbial Science Teacher Workshop, Harvard University, Cambridge, MA
    — Workshop Developer/Lecturer—Microbial Symbiosis, 2008

Harvard Museum of Natural History, Harvard University, Cambridge, MA
    — Adult education course developer/lecturer—Microbial Symbiosis,
       2008

Microbial Science Teacher Workshop, Harvard University, Cambridge, MA
    — Workshop Developer/Lecturer—Microbial Symbiosis, 2008

Science in the News Boston, Boston, MA
    — Lecturer and newsletter contributor for general public science
       outreach program, 2006

REVIEWER FOR

National Science Foundation - Division of Environmental Biology, Research Grants Council of Hong Kong, *Microbial Ecology, FEMS Microbiology Ecology, Ecology Letters, Ecology, Ecological Entomology, New Phytologist, Plant & Soil, Biological Invasions, Fungal Ecology, Mycorrhiza, Pedobiologia*

INVITED SEMINARS

2014	Yale Food Systems Symposium
2014	MIT Microbial Systems Seminar
2014	University of Nebraska, Food Science and Technology Department
2013	University of Michigan, Medical Scientist Training Program
2012	University of Wisconsin-Madison, Center for Dairy Research
2012	Science of Artisan Cheese Conference, UK Specialist Cheesemakers' Association
2009	Swarthmore College, Department of Biology
2009	Mycological Society of America Meeting, Symposium on the Conservation Biology of Fungi
2009	Point Reyes National Seashore Science Lecture Series
2007	USDA Interagency Research Forum on Invasive Species
2006	17th Annual Harvard Forest Ecology Symposium

document, since it includes all of your educational and work experiences. From this exhaustive document, you can then craft a new document that highlights areas of significance for a specific job.

In counseling PhDs who are going through this process, I first encourage them to find a sample job ad of interest. I once worked with a PhD in cognitive psychology named Julie Shapiro who used fMRIs to study brain functioning in short-term memory. Shapiro was considering jobs in data science at the time, but she did not know where to start when converting her academic-style CV to a resume. After sending her on a mission to find a job description of interest, she returned with an opening in data science at her state's Department of Education that required a PhD.

When Shapiro returned, I was pleased to see that the job description was quite long and detailed. (This will not always be the case, but do your best to uncover samples of job descriptions in fields of interest.) I suggest printing out the job description for the next step in this resume writing exercise, but you could certainly do this online if you prefer.

Exercise: Write a Resume

1. Find a job opening of interest and print it out.
2. Read through it and highlight important skills and experiences for the job.
3. Reread and list the five most salient skills or experiences required for the job.
4. Print out a copy of your CV and highlight your experiences that are pertinent.
5. Draft your new resume.

Shapiro went through this exercise to develop content for her new resume that was relevant to the position she wanted. She then had a few choices for tailoring her resume. She could have chosen to open her resume with a Summary of Qualifications, which acts as a quick description of who you are, what you've done, and the strength of your candidacy. Here is a sample:

Summary of Qualifications
- PhD-level cognitive psychologist with over 12 years of research experience
- Dedicated leader of laboratory group of 8 members
- Extensive experience in qualitative, quantitative, and behavioral research methods, including multivariate pattern analysis and linear regression analyses
- Proven communication skills, including 10 first-author publications in peer-reviewed journals and 15 oral presentations at national conferences

Shapiro could also have demonstrated these qualifications to a potential employer by creating a Skills category, which I would also recommend listing early in your resume.

Relevant Skills
- Proficient in Python, MATLAB, C++; familiarity with HTML and CSS
- Statistical analysis using R, SAS, SPSS
- Proficient with Microsoft Office products, including Excel and PowerPoint
- Experience mentoring undergraduate and graduate students
- Developed systems for maintaining paper and electronic records

The resume Shapiro created from her academic CV follows. She ended up choosing to list relevant skills, though I have seen many PhDs use a Summary of Qualifications effectively as well. Here is another example of such a summary:

Summary of Qualifications
Stem cell scientist with a proven track record of delivering high-quality, scientific results through critical analysis, effective writing, and strong communication skills; interested in applying scientific, analytical, and business skills to assist in bringing innovations to market.

Based on this example, draft your own Summary of Qualifications.

Exercise: Write Your Own Summary of Qualifications

The Summary of Qualifications section on your resume is composed of three to four brief statements that support your candidacy for a particular job.[1] In this section, you can write about your experience, credentials, values, work ethic, background, or anything that makes you the most qualified candidate for the job.

SAMPLE RESUME OF A DATA SCIENCE JOB CANDIDATE

JULIE SHAPIRO
Address
Phone
Email

Education

Boston University, Boston, MA; Ph.D., Psychology   2004–2010
Tufts University, Medford, MA; B.S., Psychology,
summa cum laude   1998–2002

Skills

- Qualitative, quantitative, and behavioral research methods, including multivariate pattern analysis, supervised machine learning (using linear support vector machines), and linear regression analyses
- Proficient in programming in Python, MATLAB, C++; familiarity with HTML and CSS
- Statistical analysis using R, SAS, SPSS
- Experience mentoring undergraduate and graduate students
- Developed systems for maintaining paper and electronic records
- Excellent oral and written communication skills

Research Experience

Postdoctoral Researcher   2010–present
Vision Sciences Lab, Harvard University, Cambridge, MA
- Led projects exploring the neural mechanisms of visual memory using univariate and multivariate functional MRI techniques
- Created new paradigms and measures to better explore the issues around visual object processing
- Used multivariate pattern analyses, and supervised machine learning to analyze where memory is stored in the human brain
- Created experiments using MATLAB and Python to test a variety of visual processes in humans
- Presented findings at small and large scale meetings and conferences
- Mentored graduate students and research associates
- Prepared yearly reports for funding and research approval agencies

- Awarded competitive grant from National Institutes of Health (NEI: Postdoc NRSA)

Graduate Student   2004–2010
Department of Psychology, Boston University, Boston, MA
- Used behavioral and fMRI methodologies to examine visual attention and memory in humans
- Performed varied statistical analyses using R and SAS
- Created experiments using Python
- Maintained small scale databases of experimental participants
- Mentored undergraduate research assistants

Research Coordinator   2002–2004
Pediatric Psychopharmacology, Massachusetts General Hospital, Boston, MA
- Coordinated multi-site study, performed structured clinical and cognitive assessments on clinical and control populations
- Maintained a database of participant information
- Served as the point person for all participant inquiries

Teaching Experience

Teaching Fellow
Harvard University, Cambridge, MA   Summer 2011
Boston University, Boston, MA   2005–2009
  Courses: Perception, Cognitive Psychology, Introduction to Psychology
  Led weekly discussion sections and review sessions, graded essays, homework, and exams, created quizzes, and tutored individual students.

Lecturer
Salem State College, Salem, MA   Spring 2009
  Course: Cognitive Psychology
  Created entire course, including course design and implementation, exam creation, grading, and one-on-one feedback and tutoring.

Academic and Professional Honors

NIH Ruth L. Kirschstein National Research   2012–present
  Service Award

Deans Fellowship, Boston University	2004–2008
Nominated for Teaching Fellow Award, Boston University	2006
B.S. awarded summa cum laude, Tufts University	2002
National Deans List	1998–2002

Selected Publications (full publication list available upon request)

Research Papers

1. Shapiro J. & Xu Y. (in preparation). The impact of distractors on the storage of VSTM in occipital and parietal cortices.
2. Shapiro J. & Xu Y. (2013). The role of transverse occipital sulcus in scene perception and its relationship to object individuation in inferior intraparietal sulcus. *Journal of Cognitive Neuroscience*, 25(10): 1711–1722.
3. Shapiro J., Michalka S.W., & Somers D.C. (2011). Shared filtering processes link attentional and visual short-term memory capacity limits. *Journal of Vision*, 11(10): article 22, http://www.journalofvision.org/content/11/10/22.full.
4. Shapiro J., & Somers, D.C. (2009). Effects of target enhancement and distractor suppression on multiple object tracking capacity. *Journal of Vision*, 9(7): 9, 1–11, http://journalofvision.org/9/7/9/.

Conference Presentations

1. Shapiro J. & Xu Y. (2014). Decoding under distraction reveals distinct occipital and parietal contributions to visual short-term memory representation. Society for Neuroscience (SfN) Annual Meeting. Washington, D.C.
2. Shapiro J. & Xu Y. (2011). Retinotopically defined parietal regions and their relationship to parietal areas involved in object individuation and identification. Vision Sciences Society Annual Meeting. Naples, FL.
3. Shapiro J., Sheremata S.L., & Somers D.C. (2008) Attentional modulations of BOLD activation in human posterior parietal cortex produced by multiple target selection and distractor suppression. SfN Annual Meeting. Washington, D.C.

Here are some questions and examples to help you come up with a strong summary statement. After reading each question and example, try to answer the question for yourself.

1. How much experience do you have in this profession, in this field, or using the required skills?

Example: Someone wanting to transition to science writing might say, "I've worked on drafting several peer-reviewed papers for the last 4 years."

Summary Statement: Four years as a writer and editor of journal articles, resulting in 3 peer-reviewed publications.

2. Imagine a close friend is talking to the hiring manager for the job you want. What would your friend say about you that would make the employer want to call you for an interview?

Example: The friend of a job hunter looking for a position conducting basic research in industry might say, "She was able to bring our group of different personalities together. I don't think anyone else could have done that as well."

Summary Statement: Proven leader in team-driven environments, identifying the skills of others and applying those to research goals.

3. How is success measured in your field of interest? How do you measure up?

Example: A physics postdoc wishing to make a move into data science might write, "Many of my professors have told me I am a quick learner."

Summary Statement: Reputation for learning new technologies quickly, including self-study and application of Python and R.

4. What is it about your personality that makes this job a good fit for you?

Example: A PhD candidate interested in curriculum development might write, "I am an empathetic listener, able to hear my students' needs and respond to them."

Summary Statement: Demonstrated empathy that translates into meeting the needs of students effectively.

5. *What personal commitments or passions do you have that would be valued by the employer?*

Example: Someone wanting to work in an environmental organization might say, "I am committed to educating people about industrial waste hazards that are endangering the environment."

Summary Statement: Strong commitment to preserving nature through education about environmental awareness.

6. *Do you have any technical, linguistic, or artistic talents that would be useful on the job?*

Example: Someone applying to a multinational corporation might write, "I can speak Spanish, Italian, and Russian."

Summary Statement: Multilingual: Spanish/English/Italian/Russian.

Now synthesize all of the points you made about yourself in this exercise and list them in your own Summary of Qualifications, either in paragraph format or by using five or six bullet points.

Using Category Headings

Another way to highlight relevant skills for employers is by tailoring category headings to the demands of a particular job description. For example, if the job description requires strong writing skills, you might create a category called "Writing Experience." In this category, you can take past experiences where you have written different types of work and describe those using language that is meaningful to the employer. Here is an example:

Writing Experience
- Conducted extensive background research on an unfamiliar topic.
- Compiled evidence germane to research question.
- Wrote overview of topic for experts in the field.
- Published work in peer-reviewed journal.

You can be quite literal in the headings you use to draw in the attention of your reader. I have seen some category headings as specific as "Quantitative Research Experience." Using this strategy will alert your reader to the fact that you are not only a potential candidate, but a well-qualified one as well.

Note every bullet point in the "Writing Experience" category above begins with a strong action verb, rather than the phrases "Responsible for" or "Duties included." These openers are weak and will not serve to illustrate what you have done in a particularly meaningful way.

When you are coming up with job descriptions for your resume, try also to quantify your work whenever possible. Mention accomplishments and outcomes. For a teaching assistant position, for example, rather than simply saying that you taught an introduction to chemistry, state that you taught an introductory course on inorganic and organic chemistry to a class of 112 students. Instead of noting that you conducted research with stem cells, explain that you conducted innovative research on cell generation from embryonic stem cells, resulting in a peer-reviewed publication.

The sample resume that follows began as a CV. It was converted to a resume for a consulting position. According to the job description, teamwork and leadership skills were critical, as were quantitative skills, so this job candidate chose to highlight all of these.

Q: International CVs versus American CVs: Is there a difference?
A: Of course. If you are applying for a position outside of the United States, you will want to look at sample CVs and resumes from that country. These can be found on the web

SAMPLE RESUME OF A CONSULTING JOB CANDIDATE

Alexandre Dumas

Professional email   Harvard University
Personal email | (cell)   Cambridge, MA 02138

EDUCATION

University Paris Denis Diderot, France, *PhD in Developmental Biology*, with Highest Honors
- PhD research fellowship (top 10%, €54k), Boehringer Ingelheim scholarship   2008
- Relevant courses: Quantitative genetics, statistics, modeling

Universities Paris Pierre and Marie Curie and **Denis Diderot**, France, *Master in Biology,* with Honors   2004
- Highest ranked French master in developmental biology

Ecole Normale Supérieure, Lyon, France; **Imperial College**, London, UK, *BS in Biology*, with Honors   2003
- Entry rank 81/1000+
- Relevant courses: Statistics, Math, Bioinformatics, Biology
- 5 months research internship in the UK, Erasmus scholarship

Joffre, Preparatory classes for the competitive exam to "Grandes Ecoles," Montpellier, France
- Math (Algebra, Probabilities), Physics, Coding, Biology, Ranked 2/31   2001

PROFESSIONAL EXPERIENCE

Harvard University, Center for Systems Biology, Cambridge, MA, 2009–Present
Postdoctoral Fellow (2009–2013), *Research Associate* (2013–Present)
- Earned the Damon Runyon postdoctoral fellowship (3% award rate nationwide, $150K)
- Designed an interdisciplinary strategy and coordinated a team to produce the first exhaustive map of a functional neuronal network
- Ran analytics on massive microscopy data sets with international university

CNRS, *Graduate Researcher*, Jacques Monod Institute, Paris, France, 2004–2008
- Awarded the Young Researcher Prize of the "Foundation Bettencourt" (16 nationwide, €25k)
- Led a collaboration with the University of X to build a quantitative model of cellular signaling, which won the lab a grant funding for 3 years

University Pierre and Marie Curie, *Teaching Fellow*, Paris, France, 2004–2007
- Co-designed and taught a laboratory class in developmental biology to 20 master's students
- Trained lab technicians and master's students on new laboratory technologies

LEADERSHIP

Illuminaria, *Editor*, Online journal and editorial marketplace, Cambridge, MA, 2013–Present
- Developed the business plan, edited literature and visual art sections; launching in the fall 2014

Harvard Graduate Consulting Club, *Event Organizer*, Cambridge, MA, 2013–Present
- Established a case practice network, led business journal clubs

Graduate Student Organization, *Vice-President*, Jacques Monod Institute, CNRS, Paris, 2005–2008
- Spearheaded career events with pharmas and startups
- Implemented an end-of-PhD fellowship for students needing extra time to finish their thesis
- Grew membership from 2 to 17 and managed a $10K budget

COMMUNICATION

- Published several scientific articles, including two in the highest-impact journals in developmental biology
- Won a first poster prize at an international conference (3 prizes for 900+ posters), UCLA, CA
- Presented research at multiple conferences and seminars in the USA, Europe and Asia

SKILLS

- **Software:** Proficient in Matlab, Mathematica, SPSS, STATA, MS Office
- **Languages:** English (fluent), French (native)
- **Interests:** Competitive open-water distance swimming (3 to 7 miles), contemporary Chinese visual arts, Certified Wine Expert: Level 2 (Boston University, 2012)

and in *Best Resumes and CVs for International Jobs: Your Passport to the Global Job Market.*[2] If you are applying for a job within the United States, you want to be sure to remove any pictures and demographic information (birthdate, marital status, or children) from your CV, as these might introduce bias in the United States but are common elements in CVs in the European Union.

Writing a Cover Letter

Drafting a cover letter is often more difficult than crafting a CV or resume because it requires the construction of a narrative, not simply a list of past positions, degrees, and accomplishments. Still, with work and lots of editing, it is possible to draft a persuasive letter to employers.

A cover letter should introduce yourself and declare your interest in the job and organization. It should highlight your job-related accomplishments and state why you are a good fit for the organization, and it should invite the employer to contact you, so make sure your contact information is highly visible.

Cover letters are typically one to one and a half pages in length, but may be longer in some disciplines and for particular job openings, such as faculty jobs. Regardless of the discipline or the job, cover letters must be tailored to the position you are applying for, must be well written, and must convince the employer of your interest in the position.

Exercise: Cover Letter Writing

Here are simple instructions for writing a cover letter, followed by further tips to get you started.

In the opening paragraph, explain why you are writing, and how or where you heard about the job.

The middle paragraphs should highlight your past accomplishments, including any activities that are relevant to the position you are applying for. This may include volunteer activities, internships, leadership in a student or postdoc organization, or other notable accomplishments. You should supply evidence that supports your candidacy for the position in a specific way. You can mention job qualifications (for example, your PhD in a particular discipline), but also personality traits or skills that you see as a good match (for example, organizational skills, event planning, or advising), as well as times you have used those skills. You will also want to describe short-term goals, including how you can contribute to the organization immediately, as well as long-term goals, and how you will fit into this organization over the years.

Throughout, you should be enthusiastic and express a sincere interest in the position. Close the letter by mentioning that you have attached or enclosed your CV or resume, and provide additional contact information. Last but not least, offer thanks in advance for the employer's interest. See the examples that follow.

Q&A on Cover Letters

Questions about these documents abound, so I will share some of those I have heard most frequently.

Q: When contacting someone in my network, should I include a cover letter?

A: In this situation, I would use a very brief email to get back in touch with a networking contact, explaining why you are writing (you are on the job market, you saw a position at

SAMPLE COVER LETTER FOR A BASIC RESEARCH POSITION

Dear Director of Recruiting:

I am writing in response to the job opening as a Scientist in Biopharmaceutics at Millennium (Job Code: BIO-22012). This position came to my attention through the Takeda website.

I am highly interested in the possibility of working as a team member with both the Millennium team and other Takeda Pharmaceuticals scientists. As part of a cooperative effort among Bayer AG Wuppertal, the University of Wuppertal, the University of Tuebingen, and the University of Jena, I learned the importance of teamwork and enjoyed working in such a setting. Working in labs in Germany and America has taught me to value different work styles and recognize the strengths of a variety of team members.

As a member of the Department of Pharmacology at the University of North Carolina at Chapel Hill (UNC), I enjoyed the opportunity to learn from my colleagues. Pursuing my project to construct artificial transcription factors, which are able to control the transcription of endogenous, p53 regulated genes, I was able to access help from scientists outside the lab. This research culminated in an AACR-AFLAC, Inc. Scholar in Training award-winning abstract and ultimately resulted in a peer-reviewed publication. Furnished with excellent facilities at UNC, I performed protein modeling with the help of the staff of the R. L. Juliano Structural Bioinformatics Core Facility. This attempt led to a spin-off project on my own, using a new approach.

My graduate education and postdoctoral training at UNC provided me with a strong background in basic research and cancer. In addition to my own research, I have been building skills in microarray technology and genomics through working with colleagues and attending seminars given by coworkers using microarray analysis. In order to keep up with the field of genomics, I audited the pharmacology course "Discovery Biology & Pharmacogenomics" at UNC in the spring semester 20XX.

I feel very confident that I could be a productive member of your company and I hope that you will consider me for a job interview. For additional information, I can be reached at 919-843-0937 or yourcandidate@med .unc.edu during the day or at 919-843-2254 in the evenings. Thank you for your consideration.

Sincerely,
Your Candidate

SAMPLE COVER LETTER FOR A POSITION IN CONSULTING

Dear:

When XX told me about Luminary Lab's unique focus on innovation systems, I instantly envisioned how my skills as a postdoctoral researcher at the University of North Carolina at Chapel Hill and my passion for problem solving could contribute to tackle the major challenges faced by Luminary Lab's clients. I am extremely motivated to make an impact as a consultant and I am available to start as soon as possible.

As a researcher, I developed curiosity, creativity, and rigorous problem-solving skills. While at Harvard, I applied an innovative strategy to decode how neuronal networks interpret environmental cues to modify behavior. I implemented analytics and cutting-edge biotechnologies, and collaborated with an interdisciplinary team to design a breakthrough, fast-tracking microscope. Our result is the first functional map of an entire neuronal network.

In parallel to my scientific work, I have sharpened my business acumen by launching "Illuminarium," an online journal and editorial market-place, by attending workshops on strategy and management, and by coordinating a case practice network.

I am convinced that my combination of creative and structured thinking, teamwork experience, and intense drive would benefit Luminary Labs and its clients. Thank you for your time and I am looking forward to hearing from you.

Sincerely,
Your Candidate

their organization, and so on), a brief reminder of what you do, and a reference to your attached resume or CV.

Q: What if I am applying to a position online that requires a resume or CV submission, but not a cover letter?

A: If there is ever a chance to upload a cover letter in addition to your CV or resume, I suggest you take it. A caveat to this suggestion is to be sure not to upload one letter and submit it for multiple positions. Each letter should be tailored for each position.

Q: How should I address the letter if I don't have a real person's name?

A: Try to get a real person's name through your contacts, online, or otherwise. If you can't find one, use a title like Hiring Manager, Director of Human Resources, or similar.

Reference Letters and Other Materials

Some organizations may request reference letters as a part of the application process. This does not happen much outside of faculty applications, as most employers will simply ask for a list of references. This list should include the name, title, organization, full address, phone, and email of at least three references. Still, you might want to have three letters in reserve. These will typically be written by your PhD advisor, postdoc advisor, and perhaps a member of your PhD committee or other collaborator or mentor in science.

Here are a few tips for getting the most out of your request for letters from references. First, ask far in advance, and ask directly whether the reference can draft a letter that will unquestionably support your candidacy. You may want to consider offering to provide letters, even if not requested. Some references may be grateful for the offer. Provide your CV and other helpful information to each reference, including a full description of the position you wish to obtain. Follow up after one month to be sure the letters have been written or are in process.

Your potential employer may request other materials besides reference letters or a reference list. Additional materials requested for academic positions include teaching statements, research statements, sample publications, a writing sample, a diversity statement, transcripts, and more.[3]

Job Search Methods

Now that you have assembled the different written pieces you need to apply for positions of interest, let's explore various job search strategies.

Network

Many job search methods exist, but *none* are as powerful as networking—and few work as well without networking. Think about infusing networking into everything you do—into your scientific collaborations, research projects, and professional meetings. Try to build relationships with people wherever you go. According to PhDs who are employed in a variety of sectors, *69 percent indicated that they found their jobs through networking.* Use this job search method regardless of what other strategies you try. For more tips and insights into the networking process, review Chapter 7.

Apply for Jobs Online

The job search strategy that seems to garner the most frustration from graduate students and postdocs on the job market is applying for jobs online. If I had a dime for every time a PhD came to me exasperated because he or she had received no response to ten, twenty, thirty, or sixty applications submitted online, I could retire today—quite comfortably. Although this job application method is free and easy, as it can be done from any lab bench, computer station, or library anywhere, it yields the poorest results in terms of contact back to the applicant.[4]

Still, employers have contacted a few people I have worked with over the years after they have submitted applications online, even without a networking contact. However, I would wager to say that number is under ten. What is much more typical of successful job applicants is that a network of contacts has been built up over time, a job ad appears online, the PhD reaches out to a contact at that employer, and a resume or CV is passed along. This is why it is so crucial to be building your network at all times.

> *Q:* I have not lined up a date for my defense yet, but I just
>   found a job posting online that I think would be a great fit for
>   me. Should I apply?
> *A:* Yes! You should *always* apply for jobs that you think are
>   a great fit for you, regardless of when they come up. You
>   never know how flexible the potential employer might be
>   about the start date, flex time, or finishing your degree.
>   Many of these can be negotiating points if the job is offered
>   to you.

You should always try to find a contact at your potential employer through your connections. It will increase the chance that someone within the organization will actually see your application. Reach out to learn more about the position and organization, even if it is a cold contact. Some employers offer a referral bonus to current employees if they refer a candidate who ends up getting hired!

Because applying for jobs online is easy and free, it is difficult not to take advantage of this strategy. Also, some fields require online applications, so it may indeed be a mandatory step in the process. My advice would be simply not to let online applications become a drain on your time and energy. Your time and energy are much better spent in developing new relationships and networking to increase the chances that you will know someone in a particular organization who could introduce you and let your resume or CV follow, rather than having your resume or CV introduce you.

> *Q:* I have a contact at a company that has a job opening online now, but the job is in a different division than my contact. Should I bother my contact about this job?
>
> *A:* Yes! Don't worry about your contact being in a different department, even if your contact is in marketing and the position is in research and development. It is still worth reaching out to ask whether your contact knows anyone in the other department so you can learn more about the position.

Attend Career Fairs

This strategy is a better one than applying for jobs online, since you will have the chance to interact with professionals in person. However, not all job fairs are created equal. You want to be sure that the event you attend is geared toward, or at the least has opportunities for, PhD-level scientists.

Things you should know when preparing to attend a career fair:

- Research the employers attending in advance.
- Practice your elevator speech.
- Bring several copies of your current resume or CV. Include a summary statement at the top, as this document will not be accompanied by a cover letter.
- Determine which organizations you want to be sure to meet with, and plan your time accordingly. (Some events will have long lines.)
- Dress professionally.
- Smile, and be enthusiastic when approaching people.

- Be prepared to talk with nonscientists and scientists alike, as some organizations will only send one or the other of these groups.
- Offer a firm—but not crushing—handshake.
- Collect business cards from everyone you meet.
- Follow up with those people you enjoyed meeting.

Also, network with organizations you might not have considered. I have met a few postdocs who have found jobs by approaching employers they were not familiar with before a career fair. Be open to new possibilities.

Use a Placement Firm or a Headhunter

Some PhDs are curious about the effectiveness of this job search strategy. If you can find a firm that assists professionals in your field with finding jobs, there is little harm in filing an application, with a few caveats. First, you should not be charged a fee. Most placement firms and headhunters are offered hefty fees by the hiring employer for finding and "placing" scientists in jobs, so the candidate—you—should not be charged. You should also know that some headhunters may contact you about positions in which you have no interest, so it is important for you to have engaged in the self-assessment process and know your skills and interests well enough to articulate them to such firms or individuals. Finally, placement firms and headhunters may facilitate introductions or share available positions with you that you were unaware of, but it is ultimately your responsibility to interview and get the job.

Find Temporary or Contract Work

A few years ago, I spoke to a recruiter who worked within a large, well-known pharmaceutical company, and he suggested two ways of breaking into an organization: through working as a temporary employee, or "temping," or through working as a postdoc in the organization first. Staffing agencies exist to place scientists and others in short-term positions. The agencies are typically hired by large firms who have an immediate need for someone to work on a project. These project-based placements can last from a few months to several years, depending on the need. This can be an effective way to break into an organization, since the hiring managers will get to know you and your work while you get to know the organizational culture and gain concrete experience. Once there, you may be considered for a full-time, permanent position, should one become available.

In addition to temping, which is typically a short assignment, you might also consider applying for a postdoctoral position in an organization of interest, which will likely be a longer-term placement. Postdoctoral fellowships in industry, for example, tend to run for two years and are more structured than temporary placements, usually offering professional development activities and training, and recruiting a sizable cohort at the same time.

An internship is another temporary position that may help you find full-time work within the organization. You may learn about openings before they are posted, and having been an intern with the organization demonstrates that you have an in-depth knowledge of their work and can interact with current employees effectively.

Work with a Career Counselor

Trying to find a job on your own is a difficult endeavor. It is time consuming and may at times be discouraging. Consider working closely with a career counselor while you are on the job market. It may be helpful for you to have someone to be accountable to, to check in with, to solicit advice from, and to boost your spirits if your search is not going well.

If you are currently a student or postdoc, you may have access to career counseling via your institution's career center or postdoctoral affairs office. Call around to see which unit offers individual counseling appointments and schedule a time that works for you. If you visit and find that the career counselor you met is not a good match, try again. It may take a few visits to find the person who is the best match for you.

If you are no longer at your institution, you should still contact the career and postdoctoral affairs offices there, in addition to the alumni office, to determine whether alumni of your institution have access to individual career counseling appointments. This is often the case, frequently through career centers, so it is worth a few phone calls to investigate. Many centers now offer Skype appointments to alumni who are all over the world, so do not hesitate if you are far from your PhD or undergraduate institution.

If you find you do not have access to career counselors through your former institutions, take advantage of free resources in your community. Many libraries offer networking events and programs on resume writing and effective interviewing. There are also many programs offered by American Job Centers (also called One Stop Career Centers), a service hosted by the Department of Labor in cities across the country.[5] You might also ask around for referrals for career coaches from friends, family, or colleagues.

Be Open to Chance Encounters

As you know, serendipity cannot be planned for, scheduled, or written into your career development plan. However, it is crucial for you to remain open to possibilities that may come your way. Is your PI inviting you to participate in a university-industry collaboration? Is your aunt insisting that you meet her neighbor, who turns out to be the chief information officer at a climate research facility? Are you sharing a room at a conference with a student whose father is a faculty member at a university in the city you want to relocate to? Keep your eyes and ears open, and be ready for fate or chance to intervene, as this is how many people find not only their first job after their degree, but also subsequent jobs, promotions, moves, relationships, and more.

CHAPTER 10

How to Interview and Negotiate

Advising PhDs about interviewing has always been a favorite activity of mine. Job seekers' tales of blunders are endless—and are great stories to tell, since they provide tools for learning how to handle oneself effectively as well as pitfalls to avoid in the interview.

Here are a few good ones:

- A PhD interviewing for a job as a medical science liaison told the interviewer about a time when he got so irritated with a vendor that he yelled at him.
- One PhD wore flip-flops to an interview—at a consulting firm in New York.
- Asked about weaknesses, a PhD interviewing for a fairly senior position in university administration went into great detail about how disorganized she was and how she had trouble finding anything on her desk. Ouch.

These scenarios may seem over the top, but they are not inexcusable. Many—perhaps even most—PhD students and postdocs have never attended an interview in a formal setting. Yet even though most PhDs have not been exposed to interviewing in the past, going into an interviewing situation unprepared is unacceptable—and preventable.

Before we launch into a more in-depth discussion of how to prepare effectively, let's review the different types of interviews you may encounter.

Types of Interviews

Regardless of the type of job you are applying for, you may be invited to interview by phone, Skype, Google chat, video conferencing, or in person. How you begin to think through and prepare for your interview depends in part on the type of interview you will be having.

Telephone Interviews

If you are actively job seeking, you need to be prepared to receive phone calls—and when you cannot talk easily, you need to let your calls go to voicemail. This means you need a professional, clearly stated message on your phone. And when you are first contacted for a telephone interview, it is most important for you to know this: *you do not have to take the telephone call right away.* You can either let the call go to voicemail, or if you do answer, follow these steps: be gracious, be enthusiastic, excuse yourself for not being able to talk right away, ask to arrange an alternate time that works for the caller, and close with gratitude for the call.

Once you have set up a time to talk, it is important for you to prepare a space free from distractions. For telephone interviews, choose a quiet space and time where and when you know you will not be interrupted. This is crucial. You don't want to be distracted by background noises. Dress for the interview. This may seem counterintuitive, but dressing formally for a telephone interview enhances your confidence and will help you to remain focused and poised during the interview. Have supporting materials in front of you. One benefit of phone interviews is that your interviewers cannot see you, giving you a chance to have supporting documents close at hand, such as your resume or CV, your responses to sample questions, employer information, and questions you have for the interviewers.

During the interview, it can be tough to keep track of participants if there is more than one interviewer on the call. One way you can try to address the right person is to draw a seating chart with names attached while the interviewers are introducing themselves. This may assist you during the call to keep people straight. Just as what you wear to your phone interview may make you more confident, smiling often during the interview will convey your genial nature and your enthusiasm for the position and organization. People can actually hear you smiling over the phone.

Close by asking about next steps. This goes for all interviews, regardless of type. Near the close, when you are invited to ask any remaining questions, ask what the next steps in the interviewing process will be. You will most often be given a timeline for what may take place next, such as in-person interviews, and when decisions about offers will be made.

Skype or Other Video Interviews

You'll need to prepare for a video interview in the same way that you would prepare for a telephone interview, with a few additional steps.

Test the software the day before to be sure it is working. If possible, do this in the quiet space you intend to use for the interview. You will need to make sure that both you and your interview space are clean and professional looking. If you do not have a space available, you might check with your career center or postdoc office, as one of these units may have private interviewing suites available.

During the video call, you may have paperwork available, but you may not want to refer to it as often as you might if you were on a voice call. Use your best judgment and try to keep distractions to a minimum. Also, be cognizant of the need to make eye contact, which can be more difficult in a video interview. You may be focused on the person on your screen, but you also need to look at the camera.

In-Person Interviews

Interviewing in person for a position may take many forms. You may be interviewing for a faculty job on a campus for two days with lots of different types of meetings. You may be interviewing in a small office with just one person. You may be interviewing with a panel of people. Regardless of the scenario, be sure to review all of the tips above, as well as the strategies that follow.

Be Kind to Everyone

While this may sound like common courtesy, I have known of PhDs who have interviewed for different jobs and have been rejected because of the way they treated the administrative assistant who greeted them. Be respectful, polite, and professional with *everyone* you meet.

Dress Professionally

This question arises frequently among PhDs, especially among scientists, for what you may wear in the lab or in your research group is usually quite informal. The safest bet for any position is a professional suit. Women may wear pants or skirts, matching jackets or different jackets. Men may wear matching or different jackets as well, as long as they look professional. I have seen some men wear ties with their suits and others choose slightly less formal attire. If you are wearing a sport coat rather than a matching suit jacket, I would wear a tie. If you are wearing a matching suit, a tie may be optional, though it does make the outfit look complete. In general, do not worry about overdressing. You may indeed be dressed more professionally than those you meet, but wearing a suit indicates that you take the job—and the interviewers—seriously, and that you are a professional. Be wary of too much cologne, perfume, jewelry, or anything that might distract from the content of your answers, or from your personality.

Be Enthusiastic When You Arrive

The first five minutes *matter*! How you approach someone impacts that person's perception of you, sometimes indelibly. Offer a firm—but not crushing—handshake and a big smile when you first walk up to your interviewer. Repeat your handshake and smile with every person you meet.

Be Aware of Your Nonverbal Signals

Nonverbal actions convey messages that are at times stronger than words. Rolling your eyes, for example, can convey boredom at best, rudeness at worst. That is an extreme example, of course, but be aware of nervous habits you might have. Do you tend to play with your hair when you are nervous? Twirl a pen in your hand? Shuffle your feet? Confidence in yourself and your abilities can be communicated through your nonverbal messages, so be aware of any anxious habits you might have, and try to avoid them during the interview.

Academic Job Interviews

Because the world of academic job interviews is different from other kinds of interviews you might encounter, I will review this process here.

Some university search committees reviewing applications for faculty positions will begin the interviewing process with telephone or video interviews. Most of these, though not all, will be conducted with more than one person. To prepare, review the tips outlined above for these two types of interviews.

Other institutions may host first-round interviews at scientific conferences. Other universities or research institutes skip these steps and ask you to visit the campus. While this is unusual, it is not unheard of, and is typically limited to top candidates.

On-campus interviewing visits can be grueling. They may take place over the course of one or two days, and are typically packed with activities that might include any of the following:

- meeting with the chair of the department, the search committee, the dean, individual faculty members, students and postdocs, or human resources representatives;

- a formal presentation of your current and future research or a "chalk talk," an informal presentation of your future plans, or teaching a class;
- a tour of the college or university, or a tour of the city or town;
- attendance at a college or university event or meals with anyone you have met in the process.

To prepare effectively for this kind of trip:

1. Ask for a detailed agenda in advance.
2. Read about each faculty member in the department, as well as those people you are scheduled to meet.
3. Get plenty of sleep.
4. Bring your talks/class lectures in several different formats (hard drive, flash drive, cloud, paper, etc.)
5. Bring several copies of your CV, and a few sample papers.
6. Bring snacks and water you can have in between meals, as you may be talking too much to eat during scheduled meals.

By the time you have reached the on-campus interview stage for faculty positions, the department is fairly certain that your work is a good match for their institution. At this point, what the interviewers tend to be assessing is *fit*. Be yourself, and be sure to get to know the other members of the community well. You need to know that the institution is a good fit for *you* as well.

Case Interviews

In consulting, there is yet another style of interviewing, and that is the case interview. In addition to a telephone or in-person interview, the case interview requires you to solve analytical or logic problems. Consulting firms use this strategy to assess your ability to think critically under pressure. You will be asked to perform computations in front of others, creating an artificially (or perhaps truly!) stressful environment. For assistance with sample case questions and strategies for preparing responses to them, see *Case in Point* by Marc Cosentino.[1] I would also recommend joining a consulting club at your institution, if possible, where case studies will be reviewed and strategies examined in a group setting.

Types of Questions

Having discussed the types of interviews you might encounter, let us now explore the types of questions you might hear. There are essentially two basic sets of questions asked during interviews: those about yourself and your work, and those about the employer you are interviewing with, or the particular job you are interviewing for (Figure 10.1).

Let's take a closer look at the questions you might hear. A full list can be found near the close of this chapter. Most of these questions will be about you. You should expect to be asked specific questions about your CV or resume, as well as some standard, open-ended

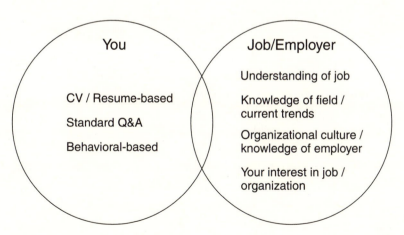

10.1. Types of interview questions

ones. Some questions may show an interest in your behavior in certain situations. Finally, you need to know what questions are *not* allowed and how to react.

Questions about Your CV or Resume

It is important for you to be prepared to discuss anything on your CV/resume. That includes work that you were only tangentially a part of, activities that took place long ago, groups that you belonged to as an undergraduate—literally *anything*. And this may be the easiest part of your interview, as you presumably know these documents very well.

Here are some sample questions:

"Why did you choose this graduate program?"
"Can you tell me about this research project?"
"I know a faculty member on your dissertation committee. How might she describe your work style?"

Standard Interview Questions

There are many stock questions that employers use during interviews to get a better sense of you and your work, such as:

"Tell me about yourself."
"What are your greatest strengths?"
"How would your peers describe you?"
"Tell me about a weakness you have."

In each case, be specific, succinct, and know that employers *always* have the following question in the back of their mind: *"What can you do for this organization?"* If you approach every answer with that question in your mind as well, you will be better able to persuade an employer that you are indeed the best candidate for the job.

Behavior-Based Questions

While standard questions about you and your CV or resume are undoubtedly helpful to employers as they seek to get to know you better, they cannot be used to assess *how* you might behave in a particular situation. For this reason, employers use behavior-based questions quite frequently. The philosophy behind these types of questions is that your past behavior is the best indicator of your future behavior.

Knowing how you reacted and handled yourself in past situations is the best evidence an employer has for how you might handle yourself on the job. Behavior-based questions, then, are situational in nature. For example, many such questions begin with "Tell me about a time when . . ."

To handle these questions effectively, use the following three-step model:

1. Describe the *situation*.
2. Explain what *actions* you took.
3. Demonstrate the *results* you achieved.

Discussing the results of your actions is important. Let's examine how to respond when faced with this question, "Tell me about a time when you disagreed with a colleague."

You might begin by describing the situation, and then move on to say that you ended up convincing your colleague to go with your decision. This statement, though, is not nearly as powerful as explaining the *result* of your actions. You might have persuaded your colleague to see things your way, but what was the end result? Did the experiment succeed, contributing data that was ultimately published, or did you end up ostracizing your colleague and others in your research group?

Focus on results in your responses. It is reasonable to share a negative result, but you must end your response on a positive note by discussing what you learned from the experience. Your potential employer needs to hear not only how you tackle certain situations, but also how adaptable you are when things do not go your way—as inevitably, they won't.

Some questions may encourage you to share a negative response and the learning outcome, such as the ever-popular question, "Tell me about a time when you failed."

Again, be specific. Do not end on a negative note, but discuss what you gained from your failure. I know a PhD in physics, for example, who failed his qualifying exams three times. What was the outcome of this repeated failure? Unwavering perseverance, which he later applied to building his own company.

Illegal Questions

From time to time, an employer may ask a question that straddles the line between inappropriate and illegal. How will you recognize these questions? And, more to the point, how will you respond?[2]

Questions related to your birthplace, ancestry, or national origin are illegal.

"How long has your family been in the United States?"
"That's an unusual name—what does it mean?"
"How did you learn to speak Chinese?"

It is, however, acceptable to ask: "Are you eligible to work in the United States?"

Questions related to your marital status, children present or planned, or pregnancy are illegal.

"Are you planning to have children?"
"What does your husband, wife, or partner do?"
"What are your child care arrangements?"

However, a potential employer can ask: "Would you be able to work a 9:00 AM to 6:00 PM schedule?" but only if this question is asked of all applicants, and a specific work schedule is a necessity for the job.

Questions related to any physical disability, your health, or your medical history are illegal.

"Are you able to use your legs at all?"
"Do you have any pre-existing health conditions?"
"Are you on any medication?"

It is acceptable for a potential employer to ask: "Can you perform the essential functions of the job, with or without reasonable accommodation?" but only if this question is asked of all applicants.

Questions related to religion or religious days observed are illegal.

"What is your religious affiliation?"
"What religious holidays do you celebrate?"
"Do you attend church or synagogue every week?"

A potential employer can ask: "Can you work on weekends?" if this question is asked of all applicants, and weekend work is a job necessity. Questions about your age are illegal.

"How old are you?"
"What year were you born?"
"I went to high school in Oakland, too—what year did you graduate?"

It is acceptable for a potential employer to ask: "Are you over the age of 18?"

Questions related to criminal records are illegal.

"Have you ever been arrested?"
"Have you ever spent a night in jail?"
"Have you ever been caught driving drunk?"

However, a potential employer can ask: "Have you ever been convicted of a crime?"

In addition, the following interview questions are illegal:

"Was your military discharge honorable or dishonorable?"
"Have you ever brought a lawsuit against an employer?"
"Have you ever filed for Workers' Compensation?"
"Have you ever been sexually harassed?"
"How much do you weigh?"
"Do you use drugs or alcohol?"

When faced with any of the illegal questions above, what should you do? It is entirely possible that the question you are asked arises out of an awkward attempt to make conversation rather than an attempt to glean forbidden information. If someone notices that you went to the same high school, they may be interested in finding out whether you were there at the same time. I once spoke to a faculty member on a search committee who wanted to know whether a candidate for a tenure-track job had any children. This was not because he wanted to discriminate against her—he wanted to share with her that the public schools where the university was located were excellent, as he was hoping to recruit her.

You have a few options when you are confronted with a question you believe to be inappropriate or illegal. First, you might answer the question directly. I would recommend this course of action only if you feel comfortable doing so. You might also consider the motivation behind the question and address the underlying concern. If someone asks whether you have a partner, they may be concerned about the possibility of you spending time away from the job. You could respond simply by saying that you have no commitments that would take you away from your work. A different response would be to ignore the question entirely and change the subject. Some people are able to evade an inappropriate question quite gracefully by using humor. Finally, you could refuse to answer the question, asserting that it is an illegal question and has nothing to do with the job at hand. I would recommend that you do this with the utmost tact, though you may still emerge as a less popular candidate than others, particularly if the interviewers were simply making small talk.

Questions about the Employer

Not all of the questions you hear will be requests for information about you. Some may be directly related to the job at hand and your interest in the position or the organization:

"Why did you choose to interview with our organization?"
"What interests you most about this position?"
"What do you know about our field?"
"What trends do you see impacting our industry?"
"Who are our chief competitors?"

Preparing responses to these questions requires research prior to the interview.

How to Prepare for an Interview

Now that you have a better understanding of the types of questions you might encounter, it is time to prepare for the interview. You will need to analyze yourself *and* your potential employer.

One way you can prepare for questions about yourself is to use the following model:

1. Read through the job description several times and highlight skills and experiences required for the job.
2. Next, go through your resume or CV and highlight experiences you have had that provide evidence of your abilities and skills that match the criteria for the job.
3. Draft a concrete example of how and when you used each skill. For example, "I developed sound research skills through my work at the National Institute for Environmental and Health Sciences investigating . . ." Or, "This year, I submitted

two articles that were accepted for publication, demonstrating my effective writing skills . . ."

4. Finally, use these anecdotes to practice answering interview questions with a friend, supervisor, colleague, or partner.

Other potential areas you need to review in advance of your interview include your dissertation, current research, and future plans. Think carefully about how your background, interests, and future goals might serve the potential employer, and be sure to frame your responses in that way. For example: "Going forward, I would like to focus on studying this particular protein, and that is why I find your research so exciting . . ."

To learn more about the employer you are interviewing with, start with a web search:

- visit the employer's homepage;
- read through their mission statement;
- study their core business, products, or services;
- learn about the history of the organization; and
- read the bios of leaders of the organization or department, as well as board members.

For academic interviews:

- look for the institutional mission and long-range academic plans;
- study the department's faculty, staff, researchers, and areas of focus;
- learn about the student population;
- read through the courses offered;
- research the current faculty and any grants or awards they have received.

Whether interviewing for academic or nonacademic positions, you can conduct research on the employees at the potential employer of interest. To find more information on those who work for the organization, you can look in several different places. Your interviewing agenda should include the names of employees you will meet, and you can search on the Internet for more information about them. LinkedIn might include the names and profiles of current employees of the organization you are interviewing with. Last but not least, you can search publication databases by the employer name to find people who work there. Although it is not necessary to know the background of every person at the organization, you can impress potential colleagues with your knowledge of different research areas, interests, or functions within the organization.

To find more information about trends and the future of the employer, read trade journals or relevant publications in their sector. If you are interviewing for a faculty position, search the archives of the *Chronicle of Higher Education* for more information on that particular college or university. If you are interviewing with a start-up that focuses on biotech, search business publications like *Boston Business Journal* or *SF Biotech*. For information about tech companies, try *Fast Company* and CNET, and if you are meeting with a think tank, look at news items related to their work in the blog "On Think Tanks: Independent Research, Ideas and Advice" (https:// onthinktanks.org), as well as any journals published by the think tank itself.

Practice Interviewing

Once you have conducted a fair amount of research on yourself and the employer, practice, practice, practice interviewing. And then practice some more. Interviewing is not a skill we develop naturally, and those who practice their responses will feel more comfortable in the interview situation and may enjoy more successful outcomes as a result of their fluency and competence. You can certainly choose to practice with a friend or partner, but most career centers, many postdoctoral affairs offices, and some alumni offices offer mock interviewing sessions for PhDs, either in person or online. Some career advisors will even offer to record the session, review it with you, and offer concrete feedback based on your performance. Take advantage of these resources, as they will serve to strengthen your chances of performing well and ultimately receiving an offer.

Exercise: Sample Interview Questions

Use this list of sample questions as you prepare for interviews. While this list is not comprehensive, it contains some of the most widely used interviewing questions, and crosses disciplines and sectors.

1. Tell me about yourself.
2. Why did you choose to interview with our organization?
3. What interests you most about this position?
4. What can you offer us?
5. What are your greatest strengths?
6. Can you name some weaknesses?
7. Why should we hire you rather than another candidate?
8. What do you know about our organization (products, services, research, or departments)?
9. What did you enjoy most about your last employment experience? Least?
10. How do you think your advisor or supervisor would describe you?
11. Why do you want a job for which you seem to be overqualified? (if applicable)
12. What skills have you developed through your graduate studies and postdoc that relate to this job?
13. What are your salary expectations? (Answer with a question: "What does the typical range look like for someone with my experience?")
14. Why are you applying for positions outside of academic settings? (if applicable)

Behavior-Based Questions

15. Describe a time when you had difficulty working with a professor, advisor, supervisor, or coworker. How did you handle it?

16. Give an example of a time when someone criticized your work in front of others. How did you respond?

17. Give an example of a time when you sold your supervisor on an idea or concept. How did you proceed? What was the result?

18. Describe the system you use for keeping track of multiple projects. How do you track your progress so that you can meet deadlines? How do you stay focused?

19. Tell me about a time when you came up with an innovative solution to a challenge your lab or group was facing. What was the challenge? What role did you play?

20. Describe a specific problem you solved for your employer or professor. How did you approach the problem? What role did others play? What was the outcome?

21. Describe a time when you got coworkers who dislike each other to work together. How did you accomplish this? What was the outcome?

22. Tell me about a time when you failed to meet a deadline. What did you fail to do? What were the repercussions? What did you learn?

23. Describe a time when you put your needs aside to help a coworker understand a task. How did you assist them? What was the result?

24. Describe two specific goals you set for yourself and how successful you were in meeting them. What factors led to your success in meeting your goals?

Questions Asked during Academic Interviews

1. Describe your current research.

2. Why did you choose to focus on this area?

3. What will your next research project be? Are you planning to make any future changes to your current project?

4. Describe your philosophy of teaching.

5. How do you motivate students?

6. Describe a course you have taught in the past and how you evaluated the students' learning.

7. How would you teach this (introductory level, intermediate, advanced level) course? What primary and secondary texts would you choose?

8. How have you used technology in the classroom?

9. How would you increase enrollment in this major?

10. Describe your ideal course. What does the syllabus look like? What texts would you envision using?

11. Why are you interested in this college or university?

12. Tell me where you see yourself in five, ten, or twenty years.

You should also be ready with your own questions for the employer. If an employer asks whether you have any questions at the beginning, middle, or end of your interview, your answer should always be "yes." If you say no, it reflects poorly on your level of interest in the employer or position. Here are a few sample questions:

1. Can you describe the primary responsibilities of this position?

2. What does a typical day look like?

3. What is the largest single problem facing your staff (department) right now?

4. May I speak with the last person who held this position?

5. What is the usual time frame for promotions?

6. What do you like best about your job or this organization?

7. Has there been much turnover in this area?

8. How is performance evaluated?

9. What qualities are you looking for in the candidate who fills this position?

10. What is the management style of the person who will serve as my supervisor?

11. Is there a lot of teamwork? Project work?
12. What is life like in this city or town?
13. What are the next steps? When should I expect to hear from you?

How to Close an Interview and Follow Up

As your interview draws to a close, try to summarize themes you hope to leave with the employer, and be sure to ask any questions not answered for you in the course of your time together. You might mention any key items not covered that are relevant to the position.

State your continued interest and ask the employer what the next steps for the search might be. Employers will typically share the timeline for a search with each candidate, letting you know when to expect a response. Before leaving, request a business card or contact information from your main contact for the day.

Following an interview, it is critical for you to send a thank-you note within two days. Print or email is fine, as long as your response is sent quickly and imparts a professional and gracious tone. An added bonus of sending a thank-you note is the opportunity to remind the interviewer of why you are the best person for the job. You might say something like this:

Dear Jane:

Thank you for the time you spent with me today describing the current and future research plans of NextGen Inc. It was a pleasure to talk with you, and I am excited about the opportunity to apply my skills and experience in molecular genetics to advance your team's initiatives.

Thank you once again, and have a wonderful weekend.

All the best,

Joanna

Once an employer has made a decision on the best candidate for the position, you may receive a phone call or email with the news. If the news is negative, be sure to remain gracious, and ask for feedback if you feel comfortable doing so. Some employers are willing to share specifics with you that will enhance your interviewing skills and strengthen your chances for future interviews. You might also ask whether you may stay in touch with the employer, if you had a positive interaction. While you may not have been the right fit for this particular position, you may be a good fit for the organization overall and want to remain in contact in the future.

If you do not hear back from the employer in the time stated at the end of the interview ("by close of business tomorrow," "by Friday," "by the end of next week"), it is entirely appropriate to email or call to follow up. When you do reach out, there is just one question that I would recommend asking: "What is the timeline for your decision?" Use this wording, rather than ask about your candidacy in particular. This takes the onus off of the employer, and will generally yield information that you can use, such as: "the search committee hasn't met yet," "our vice president is out of the country right now and won't be back until next Tuesday," or "we won't have an answer until next month."

After the initial contact and timeline given has passed, it is appropriate to follow up again two weeks later if you still have not heard from the employer. Searches take time, and often are not the top priority at an organization, so be patient, if you can.

Figure 10.2 includes a few final tips on the interview process.

Negotiating an Offer

If you are contacted after an interview and the response is positive, you may be invited back for another interview, or you may receive an offer. When you receive this call or email, I would suggest the following: be gracious, thank the employer for the offer, request the offer in writing, and ask for some time to respond. This will give you the opportunity to gather your thoughts and questions, conduct some research, and prepare to negotiate.

INTERVIEWS

DO	DON'T
• Research the department or organization before the interview	• Arrive late!
• Review sample questions and practice your answers	• Accept a formal interview to "practice" your interviewing skills
• Schedule a mock interview with a career counselor	• Ask about salary during the initial interview
• Bring extra CVs or resumes to the interview	• Ask about a position for your partner—at least not initially, and only in some circumstances
• Dress professionally	• Argue with the interviewer
• Bring a list of questions you have for the interviewers	• Volunteer negative information
• Send a thank-you letter as soon as possible	• Continue to interview after you have accepted another position
• When discussing salary, talk in terms of a range, not a single figure	

10.2. Interview dos and don'ts

Questions to consider while reviewing the offer might include:

• Is the salary appropriate for the position and for your experience and skill level? Consider well the cost of living in a particular geographic location.
• Does the position include benefits?
• Does the offer meet other criteria important to you?

The offer may be entirely appropriate and reasonable, and if you feel satisfied with it, you may accept without negotiating.

If you would like to engage in the negotiation process, you will need to do some research and prioritizing. Salary.com is a great first stop to understand starting salaries for particular positions in a specific geographic region. This site displays salary search results as a range, including the mean salary for a given position, and brief descriptions of positions, should you be unsure of the particular title match for the job you have been offered.

For specific types of positions or disciplines, you may seek other data sets, such as the American Association of University Professors (AAUP) Salary Survey, the American Chemical Society Salary Survey, or the Annual Survey of the Mathematical Sciences.[3] Once you have a sense of the typical salary range for the position, you might think also about aspects of the job that are important to you, such as the flexibility of the start date, what work hours are expected, or whether or not there is child care available. It is critical to know your priorities and to understand clearly which points are—and which are not—negotiable. Use Figure 10.3 to consider and rank your priorities. Some points may be negotiable, others

ISSUE	RANK
Salary	
Benefits (medical, dental, retirement, life insurance)	
Start date	
Vacation time	
Moving expenses	
Opportunity for advancement	
Supervisory responsibilities	
Child care on site (if applicable)	
Lab supplies and yearly budget	
Professional development support (training and conferences)	
Administrative support	
Computing needs	

10.3. Ranking your priorities in negotiations

ISSUE	RANK
Start-up funds	
Teaching load	
Summer support	
Graduate assistants	
Length of contract	
Tenure clock	
Grant writing expectations	

10.4. Ranking your priorities in negotiations for faculty positions

may not, but all represent value to you. Figure 10.4 notes some additional considerations for negotiating a faculty position.

Once you have identified your priorities, you will be more aware of where you are willing to make concessions and where you intend to hold firm. When you are ready to engage in the negotiation, set up a time to talk by phone with the person who extended the offer to you. Take the following steps during this conversation:

1. Be enthusiastic and reiterate your interest in the position.
2. Remind the person of why you think this is a good fit for you.
3. Share your concerns about an aspect of the offer that you would like to negotiate.
4. Present your request clearly, stated as an open-ended question.
5. Be prepared to make trade-offs across issues.
6. Be creative in developing a solution that will benefit both you AND the employer.

Chances are the person you negotiate with will not be able to provide an answer to your requests immediately. Typically, hiring managers must take negotiated offers back to their own managers for approval and reply to you later with the results.

Once the negotiation has concluded, request a new offer letter outlining all of the details you discussed. You will then have some length of time to respond. It is appropriate to ask for an extension if you need more time, but do not miss your final deadline.

If you decide to decline the final offer, call first and then and follow up with a letter. If you are considering multiple offers, keep all parties informed on the status of other applications. You can use your leverage to ask an institution to match an offer, but *only* if you intend to accept it.

Remember throughout the negotiation process that your ancillary goal is to build goodwill, as you will presumably be in this position for some length of time. Be sure to keep conversations cordial, rather than adversarial, and try to meet the needs of your potential employer as well as your own.

Conclusion:
Making a Successful Transition

Congratulations—you have landed a job! If you are moving into a faculty position, chances are you have been coached by other faculty members through the interviewing and negotiating stages and are now prepared to tackle the challenges inherent in launching an academic career.

If you are moving into one of the myriad choices available to PhDs discussed in Part Two, you will also need to be prepared to encounter the complexities of any work environment: personalities, expectations, diplomacy, performance, and outcomes. It is your job to learn about the culture of the environment you are entering and to handle changes with grace. However, if you have never worked outside of the academic realm, this may prove difficult.

You may even experience a crisis of identity. You may be thrilled about your new opportunity, but you may be separated from your scientific research for the first time. How you manage this crisis will determine your early success in your new role, regardless of the sector or career field you have chosen.

To help you manage the transition from PhD student or postdoctoral trainee to full-time, permanent employee, let's examine some general differences between academe and the rest of the working world.

Cultural Differences at Work

Scientists and engineers are clearly connected to research groups of one size or another, but in terms of your day-to-day work, you may be accustomed to working alone on projects. You may have been part of a large laboratory but responsible for a small piece of research that kept you busily working on your own, checking in perhaps at weekly lab meetings.

In contrast, most of the working world is built on teams. Whether you have accepted or intend to move into a research position in industry, a teaching position at a private school, or an analyst position in a consulting firm, you need to expect to participate actively in the workings of your immediate team. You will also need to recognize the structure of your organization and other teams, departments, or units within it.

In a similar vein, you may have possessed a great deal of freedom to set your own agenda as a postdoc or graduate student, but now you may be asked to work toward the goals of the organization at large. Hierarchies exist everywhere, and it is important for you to follow the lead of your supervisor, others on your team, and the organization itself as you plan your work and set your own goals. This may be quite difficult for you to accept, since you may be accustomed to complete autonomy with regard to your own work. Still, you must adhere to the culture at your new workplace in order to succeed there. The priorities of the organization may or may not align with yours, and it will be necessary for you to accept the differences therein.

In your research, you may have had the time—some might say the luxury—of exploring a research question at your own pace. You may have been invited by your advisor or department to explore this question for an indefinite period, since there may have been no organizational or group pressure for you to produce a publication. This tends not to be the case at most organizations. Whether you are engaged at a for-profit enterprise, a nonprofit, or a government agency, you will be asked to adhere to organizational goals and to let go of projects when necessary. This can be a struggle for some PhDs, since the pursuit of knowledge for its own sake has likely been a priority until now. However, PhD scientists need to learn how to balance their quest for knowledge and commitment to the process with an understanding of how organizations function and how to change directions quickly. You may also find that you need to acquire some business acumen.

As we think about cultural differences, we should take note of the informality of most college and university environments. Some PhD scientists and engineers coming out of their training period may be in the habit of wearing shorts and t-shirts to work, arriving at the lab in the late morning, playing volleyball over lunch, going home for dinner, and then returning to the lab late in the evening to finish their work. Scientists may also be comfortable swapping materials and resources with friends in other labs without formal documentation or alerting the lab manager.

In terms of dress, time, and sharing resources, as well as many other aspects of culture, it is critical to respect the ethos of your new work environment. Are the other scientists wearing t-shirts and shorts? Or do they come to work in dress pants and button-down shirts? Does your team arrive at 11 am or 9 am? Or are people on flexible work schedules, arriving when they can but always staying at work for eight hours? What is the accepted protocol for sharing resources? How do people tend to communicate? Is calling better, or does your team work primarily by email? Be sure to learn the expected cultural norms of your workplace—and if you are unsure, ask. It is much better to feel slightly awkward by asking a coworker a question than to be reprimanded by your boss for not adhering to a cultural norm.

Acquainting yourself with the norms and practices of your new employer is crucial, according to Amy Rawls, director of human resources at Research Square. When hiring, "I have to be persuaded that the applicant is a good fit for the organization," Rawls says, noting that half of a hiring decision is "culture fit." "In fact," she adds, "we describe our culture right on the Careers page."[1]

How to Transition with Grace

There are a number of steps you can take, and missteps to avoid, that will ease your way into the working world.

Do not disparage your PhD experience.   In your new place of employment, you may be surrounded by other PhD scientists, or you may be the sole PhD on your team. In either situation, you may be asked why you chose to leave academe. I encourage you to view this as an opportunity to share what you learned during your education, as well as how interested you are in embarking on this new adventure. This is *not* an invitation for you to list the many things you disliked about academe, injustices you suffered at the hands of your faculty advisor, reasons your papers were not published in better journals, and so on. This is your chance to embrace the culture of your new workplace and share with new colleagues what you would like to accomplish.

Ensure open communication.   Request a meeting with your new supervisor shortly after you arrive, if one has not been arranged previously. Be candid during this meeting, asking directly what concrete results your supervisor expects to see from you in the first month, three months, six months, and first year. Think back to your Individual Development Plan, and draft something similar with your new boss. This process may already exist at your workplace in a more formal way, but you can always draft a list of goals for yourself, both long and short term, that will keep you on track and keep the lines of communication open between you and your new supervisor. Once your goals have been drafted, you might also ask to meet with your new boss at regular intervals. Again, your workplace may already have an established means of evaluating your performance throughout the year, in which case you should follow the protocol, but if not, be sure to arrange for periodic check-ins, as this will also provide an opportunity for you to solicit feedback on your work.

Get to know others on your team—and off. Take part in group functions, eating lunch with your new team or group, but also get to know people across the organization. Resist the temptation to eat at your desk or stay at your bench during this social time. It is vital for you to build relationships at your new workplace. If there are group outings, join them as often as you are able, since this will allow you not only to get to know coworkers better, but will strengthen the group's ability to work as a team.

Be ready to learn. You may be given a project in your new role that requires you to learn a new skill, study an unfamiliar topic, or give a talk on someone else's work. Be ready, willing, and vocal about your ability to learn quickly. This is a skill that most PhDs possess, and that employers across all sectors value. Make sure to share this trait with others as opportunities arise, and you will be seen as a team player and potential leader, depending on the project and circumstance.

Build a support network. It may be helpful to connect with people who are where you want to be—fully transitioned into your new profession. Think not only about reaching out to people at your new place of work, though this can be a source; think also about reaching out to people in professional societies. Become a member and join committees. Volunteer your time to learn all that you can about the growth of the profession, the culture, and the people. In the process you may identify new mentors.

Allow yourself time to make the transition. Because you are moving from one stage of your life to another, give yourself a chance to adapt to your new identity. This adaptation includes recognition of the departure of your old identity. Take the time you need to grieve the loss of this identity—the graduate student who was involved in various committees across campus, the postdoc who was always jumping in to assist undergrads in learning new techniques, the trainee who was the last one to leave every happy hour. Whether you had the time of your life during your graduate work and postdoctoral training or found it arduous and could not wait to be finished, you need to take some mental time and space to recognize that this change is a substantial one. You will need time to adjust to your new life. Take care of yourself, see old friends, and engage in activities outside of work that make you happy. Seek a therapist if you find you are no longer enjoying your time at work or outside of work. Although adapting to a new work identity is a considerable challenge, it is not without its rewards. Identify and focus on those parts of your new job that bring you joy, and the transition will be easier for you.

Give back to the community.   Just as you had the pleasure of connecting with professionals in different fields while you were going through the process of finding a career, you can do the same for those PhDs still engaged in the decision-making process. Make yourself available through your past institutions as an alumni contact for current students. Serve as a mentor to a PhD student at an institution close by. Volunteer at a local career center to talk about your career and the path you took. Adjust your settings on LinkedIn so the public can see your profile. Contribute to online discussions of careers for PhDs or start a blog about the process. There are multiple ways you can give back to the community, and this kind of reciprocity is what keeps exciting opportunities available to future PhDs. Amy Rawls at Research Square says it best:

> Take some time with your decision. Finding the right environment can be so empowering. Follow what is most fulfilling to you and have a good time with it. That will make you a great candidate for the next great thing.[2]

Finding a career that fits who you are and what you enjoy is worth the effort. Think carefully about your options and your preferences, spend time on your search, and reach out to others. Remember that you are the architect of your own career, and you, like the majority of PhD scientists, *can* find satisfaction in your work.

Appendix A. Identifying Career Pathways
for PhDs in Science

Appendix B. Resources for PhDs

Notes

References

Acknowledgments

Index

Identifying Career Pathways for PhDs in Science[1]

Materials and Methods

Ethics Statement

This study was conducted in compliance with and approved by the Harvard University Committee on the Use of Human Subjects, IRB#15–0063. Participants in the study were ensured confidentiality. Each participant reviewed a consent form prior to taking part in an online survey and clicked on an embedded link to demonstrate consent and proceed with the survey. All of the data in this study have been anonymized.

Data Collection

The population taking part in this study included graduates who had earned a PhD in any scientific discipline, including the life sciences, physical sciences, mathematics, engineering, and social sciences at any institution worldwide between 2004–2014, and who had ever worked, trained, or studied in the United States.

Participants for this study were identified in three phases. Phase I was a qualitative phase using both convenience sampling and purposive sampling to identify participants who met the criteria above, and who were willing to take part in a structured, sixty-minute interview.[2] Five participants were recruited from former postdocs of the research universities where the study author was once employed, and through searches on LinkedIn affinity groups for PhDs interested in "alternative" (nonacademic) careers. Once identified, the participants were interviewed individually by phone about their training, educational background, current employment, and skills required to be successful. Their responses were used in part to create a survey on career choice, whose questions were also derived from surveys constructed by Davis in 2003, Sauermann and Roach in 2012, and Gibbs and Griffin for their study in 2013.[3] The new survey instrument included similar questions on education and training background, career goals, and any changes in career goals over time, but differed from past instruments in its focus on the current employment situation of the PhD, including job title, employer, sector, geographic location, and skills analysis.

Using data gathered in Phase I, as well as former surveys of this population, a brief, online survey was constructed. The new survey instrument contained embedded cognitive interview questions at the end of the questionnaire to allow for more rapid analysis of survey items. Limitations using this method of cognitive interviewing included the inability to conduct follow-up probing with respondents, and a limit to the number of items that can be analyzed this way.

In Phase II, the pilot quantitative phase, the newly constructed survey was distributed to twenty PhD-level scientists and engineers who are first-order contacts of the study author through LinkedIn who met the criteria listed above. Participants in this phase field tested the survey and provided feedback on the items through the embedded probing questions at the close of the survey.

The survey was then revised according to feedback gathered from the embedded probing questions, and Phase III consisted of the full-scale launch of the survey instrument to PhDs all over the world.[4]

Participants were invited to participate through forty-three LinkedIn affinity groups focused on PhD-level science, social science, and engineering professionals. In addition to this recruitment method, the National Postdoctoral Association (NPA) agreed to send a message from the study author to all members who attended the NPA Annual Meeting in March 2015. Invitations were also sent to Graduate Career Consortium members (over 200 academic institutions) to send to their PhD alumni. Finally, participants were recruited via snowball sampling and asked to share the survey instrument and study goals with all eligible groups and peers.[5]

Multiple reminders and different wording were used to increase response rates. In addition, participants were invited to enter a drawing for a small number of prizes (five awards at $100 each). Using both dynamic and static design features contributed to an increase in the overall response rate.[6] Unfortunately, given the sampling techniques used in the study, personalization of each invitation to participate was not possible.

All surveys were conducted online, using the software program Qualtrics (www.qualtrics.com). The number of respondents who began the final survey in Phase III was 11,076. The number of complete responses to the survey was 8,099, representing a 27 percent dropout rate.

Measures

The survey instrument constructed took approximately fifteen minutes to complete and included questions about career interests, activities, current employment, and motivations for career choices.

The instrument consisted of four major sections, including Education, Postdoctoral Training, Employment, and Demographics. Sample questions from each section are below.

Education
- From which institution did you receive your doctorate?
- In what field/discipline/academic program did you complete your doctorate?

- At the start of your doctoral program, which of the following career goals did you aspire to?
- Please indicate your level of agreement with the following statement: "I developed/continue to develop this skill during my doctoral program."
  - Discipline-specific knowledge
  - Ability to gather and interpret information
  - Ability to analyze data
  - Ability to manage a project
  - Oral communication skills
  - Written communication skills
  - Ability to work on a team
  - Ability to make decisions and solve problems
  - Ability to manage others
  - Creativity/innovative thinking
  - Time management
  - Ability to set a vision and goals
  - Career planning and awareness skills
  - Ability to learn quickly
  - Ability to work with people outside the organization

Respondents rated each skill on a five-point Likert-type scale from "strongly disagree" to "strongly agree."

Postdoctoral Training
- Which institution(s) employed you as a postdoc?
- How many different postdoctoral appointments have you held?
- What is the total number of years you have been or were a postdoc, across all locations and research institutes?
- Please indicate your level of agreement with the following statement for each skill listed below: "I developed/continue to develop this skill during my most recent postdoc."
  - Discipline-specific knowledge
  - Ability to gather and interpret information

- Ability to analyze data
- Ability to manage a project
- Oral communication skills
- Written communication skills
- Ability to work on a team
- Ability to make decisions and solve problems
- Ability to manage others
- Creativity/innovative thinking
- Time management
- Ability to set a vision and goals
- Career planning and awareness skills
- Ability to learn quickly
- Ability to work with people outside the organization

Respondents rated each skill on a five-point Likert-type scale from "strongly disagree" to "strongly agree."

Employment
- Who is/are your current employers?
- Please select the sector that best describes your current employer.
- What is your current job title?
- What are the primary activities of your current position?
- Is a PhD required or preferred for your current position?
- Is postdoctoral training required or preferred for your current position?
- Which skills are important for success in your current position?
  - Discipline-specific knowledge
  - Ability to gather and interpret information
  - Ability to analyze data
  - Ability to manage a project
  - Oral communication skills
  - Written communication skills
  - Ability to work on a team

- Ability to make decisions and solve problems
- Ability to manage others
- Creativity/innovative thinking
- Time management
- Ability to set a vision and goals
- Career planning and awareness skills
- Ability to learn quickly
- Ability to work with people outside the organization

Respondents rated each skill on a five-point Likert-type scale from "not at all important" to "extremely important."

Limitations

One limitation of the sampling methods used is sample selection bias, since these methods largely assume online activity through the professional networking site LinkedIn. Furthermore, this sampling method may miss PhDs who are unemployed or not seeking employment and thereby less active on electronic or social media channels.

Resources for PhDs

This list is not intended to be comprehensive, but merely a sample of the vast resources available to PhDs searching for a career.

General Career Resources

myIDP. Free web resource on career development for scientists, including self-assessment exercises and descriptions of a variety of occupations. http://myidp.sciencecareers.org/

National Postdoctoral Association. The NPA webpage contains articles on postdoc policy as well as a bevy of career-related resources. If your institution is a sustaining member of the NPA, you can register for a free membership using your institutional email address. http://www.nationalpostdoc.org/

Fellowships

American Association for the Advancement of Science (AAAS). Fellowships. http://www.aaas.org/page/fellowships

RAND Corporation. Internships and Fellowships. http://www.rand.org/about/edu_op.html US Office of Personnel Management. Presidential Management Fellowship Program. http://www.pmf.gov/the-opportunity/pmf-stem.aspx

Professional Societies

Association of University Technology Managers (AUTM). Career Center. http://careercenter.autm.net/home/index.cfm?site_id=3719

Career Exploration Resources

Science. Career Profiles. Website containing hundreds of articles on career options in science. http://sciencecareers.sciencemag.org/career _magazine/career-profiles

Science. "The World's Most Comprehensive Resource for Scientists Transitioning to Industry," July 3, 2014. http://sciencecareers .sciencemag.org/career_magazine/previous_issues/articles/2014_07 _03/caredit.a1400170

The New York Academy of Sciences (NYAS). Organizes many seminars, workshops, and courses. http://www.nyas.org/default.aspx

The New York Academy of Sciences: Science Alliance. This webpage contains video interviews with individuals with science backgrounds in a variety of careers (research and nonresearch). http://www.nyas .org/WhatWeDo/ScienceAlliance/Careers.aspx?tid=4ac9a42b -60f2–4391–87a0-f81535a6a41d

Skills Listings

University of Michigan Student Life: Career Center. "PhD Transferable Skills." http://careercenter.umich.edu/article/phd-transferable-skills

Skill Building Resources

The Jackson Laboratory. Courses and conferences for scientists in Bar Harbor, ME, and Farmington, CT. http://www.jax.org

Cold Spring Harbor Laboratory, Long Island, NY. Courses for students and postdocs on particular scientific research focus areas. http:// meetings.cshl.edu/courses.html

The Marine Biology Laboratory, Woods Hole, MA. Summer courses for students and postdocs, including microscopy. http://hermes.mbl.edu /education/courses/summer/index.html

World Science Journalism Federation. Science journalism course (online and free). http://www.wfsj.org/course/

The Solution Lab. Life science consulting: a nonprofit dedicated to educating graduate students and postdocs about healthcare and pharma consulting through workshops, seminars, and volunteer experiences. http://thesolutionlab.org/

The Entrepreneurship Lab, New York, NY. For life science students and postdocs interested in gaining practical entrepreneurial experience. http://elabnyc.com/

Higher Education Jobs

The Chronicle of Higher Education. Articles on the latest happenings and trends in higher education and a searchable database of job opportunities in academia, including teaching positions and administrative positions. http://chronicle.com/

Inside Higher Ed. Trends in higher education and a searchable database of job opportunities in academia. http://www.insidehighered.com /#sthash.aLurJTGi.dpbs

HigherEdJobs. A searchable database of job opportunities in academia. https://www.higheredjobs.com/search/default.cfm

Higher Education Recruitment Consortium (HERC). A searchable database of job opportunities in academia. http://www.hercjobs.org/

Nonprofit or Government

Chronicle of Philanthropy. Maintains a searchable database for jobs in the nonprofit sector at many foundations and nongovernmental organizations. http://philanthropy.com/section/Home/172/

Idealist. A searchable database of volunteer opportunities as well as jobs available at nonprofits. http://www.idealist.org/

USAJobs. A searchable database of jobs in the government sector. https://www.usajobs.gov/

US Department of State. Virtual Student Foreign Service (VSFS). Internships for college students available by application in fifteen government agencies. http://www.state.gov/vsfs/

Fulbright Postdoctoral Scholar Awards. http://www.cies.org/program /postdoc

American Society for Biochemistry and Molecular Biology (ASBMB) Hill Day. An opportunity for science researchers to learn about science advocacy and government. http://www.asbmb.org/Advocacy /advocacy.aspx?id=13812

Data Science

Insight Data Science Fellows Program. A seven-week postdoctoral
training fellowship for academics interested in data science. http://
insightdatascience.com/
General Assembly. Offers a twelve-week immersive course in data science
in Boston, MA. https://generalassemb.ly/education/data-science
-immersive

Business

Xconomy. Business, life sciences, and technology news. http://www
.xconomy.com/
Vault.com. Rankings and reviews of industries, professions, and industry
leaders.
Wet Feet Guides. Guides to different professions within business.
https://www.wetfeet.com/guides

Mini-MBA, Bioscience, or Industry-Focused Courses

Rutgers Mini-MBA, BioPharma Innovation. http://www.business.rutgers
.edu/executive-education/programs/mini-mba-biopharma-innovation
Harvard Graduate Business Club, Mini-MBA Program: http://
harvardgraduatebusinessclub.com/?page_id=287
SUNY Stony Brook, Fundamentals of the Bioscience Industry. http://fobip
.org/
Keck Graduate Institute. Managing Science in the Biotech Industry.
Intensive summer course for students and postdocs. http://www.kgi
.edu/academic-programs/summer-intensive-program-for-post-docs.html

Online Courses

American Chemical Society Online Courses. http://proed.acs.org/course
-catalog/categories/professional-development/
Harvard Business Review Online Courses. http://hbr.org/store
/landing/courses
Harawar, Devindra. "The Seven Best Ways to Learn How to Code."
VentureBeat, October 31, 2013. Lists resources for coding courses.
http://venturebeat.com/2013/10/31/the-7-best-ways-to-learn-how-to
-code/

Nisen, Max. "Fifteen Free Online Courses that Are Actually Worth Your Time." *BusinessInsider,* September 27, 2013. http://www.businessinsider.com/best-online-courses-to-take-2013–9

Teaching

Perdue, Sarah. "How to Get Teaching Experience that Will Help You Land a Job." *ASBMB Today,* December 2013. http://www.asbmb.org/asbmbtoday/asbmbtoday_article.aspx?id=49765

AdjunctNation.com: News, Opinion, Analysis and More for the Adjunct Faculty Nation. Searchable job database. http://www.adjunctnation.com/

Cell Motion Laboratories, Bio Bus: http://biobus.org/

Science Writing

Elsevier, "How to Become a Reviewer," March 1, 2012. https://www.elsevier.com/reviewers-update/story/career-tips-and-advice/how-to-become-a-reviewer

Council for the Advancement of Science Writing, "How Do I Get Started in Science Writing? A Guide to Careers in Science Writing." http://casw.org/casw/how-do-i-get-started-science-writing

Council for the Advancement of Science Writing Fellowships and Awards: http://casw.org/casw/fellowships-and-awards

National Association of Science Writers. http://www.nasw.org/

Nascent Medical, LLC. Medical Writer Resources, The HITTList. A free subscription email with numerous medical writing jobs, including many which could be performed part time as a freelancer. Course and book series on what you need to know about careers in medical and science writing. http://www.hittmedicalwriting.com/hmw/the-hittlist/

Elsevier Publishing Campus. Career resources webpage for early-stage researcher, with a particular focus on publishing. http://www.elsevier.com/early-career-researchers/home

Notes

Preface

1. National Science Foundation 2013b.
2. See Weissmann 2013; Vastag 2012; Lametti 2012.
3. "Are 'Alternative' PhD careers still the alternative?" 2013.
4. National Science Foundation 2013c.

Introduction

1. Over the past decade, the number of engineering doctorates earned has risen 47 percent, and the number of doctorates earned in the biological sciences has risen 49 percent. National Science Foundation 2000–2009.
2. See http://nsf.gov/statistics/doctorates for estimates of science and engineering doctorates awarded between 1985 and 2014, and http://nsf.gov/statistics/2016/nsb20161/uploads/1/8/at05-13.pdf, both accessed April 4, 2016, for science and engineering doctorate holders employed in academia by type of position, including full-time faculty positions.
3. Iasevoli 2015.
4. Beginning with a report from the National Academies, *Reshaping the Graduate Education of Scientists and Engineers,* that recommended better collection and dissemination of career outcomes data for PhDs in science and engineering, myriad reports have surfaced and echoed the sentiments of that original work. These include:

   1998: Committee on Graduate Education Report and Recommendations—Association of American Universities
   1998: Trends in the Early Careers of Life Scientists—National Academies

2000: Addressing the Nation's Changing Needs for Biomedical and Behavioral Scientists—National Academies

2000: Enhancing the Postdoctoral Experience for Scientists and Engineers: A Guide for Postdoctoral Scholars, Advisors, Institutions, Funding Organizations, and Disciplinary Societies—National Academies

2005: The Responsive PhD: Innovations in US Doctoral Education—Woodrow Wilson National Fellowship Foundation

2007: Rising above the Gathering Storm: Energizing and Employing America for a Brighter Economic Future—National Academies

2010: The Path Forward: The Future of Graduate Education in the United States—Council of Graduate Schools (CGS)

This list is not intended to be comprehensive, but merely to show the breadth of organizations discussing the same issue for more than a decade. In 2012, a watershed year for information about the career trajectories of the nation's PhD scientists, three reports further expanded our understanding: "Pathways through Graduate School and into Careers," by the Council of Graduate Schools; "Research Universities and the Future of America: Ten Breakthrough Actions Vital to Our Nation's Prosperity and Security," from the National Academies; and "Biomedical Research Workforce Working Group Final Report," from the National Institutes of Health (NIH).

The National Academies and the NIH reports both discussed the oversupply of PhD-trained scientists and engineers, the simultaneous dearth of faculty jobs, and the impact that the dismal faculty market was having on the desire of young people in the United States to train in science. The system, as it stood in 2012, created a strong disincentive to American college graduates to enroll in doctoral programs.

5. Selfa and Proudfoot 2014.
6. Sinche, "Identifying Career Pathways for PhDs in Science," online survey, N=3,335, May 2015.
7. See Young 2011.

1. How to Connect Your Interests to Careers

1. See Holland 1997 for more information concerning person-environment interaction.
2. If you are trying to locate career counselors or advisors in your area or at your institution, contact one of the two leading organizations of professionals who focus on PhDs and postdocs: the Graduate Career Consortium (http://gradcareerconsortium.org/) and the National Postdoctoral Association (http://www.nationalpostdoc.org/).

3. See "The History of the MBTI Assessment," https://www.opp.com/en /tools/MBTI/Myers-Briggs-history, accessed February 4, 2015.
4. "MyIDP Individual Development Plan," http://myidp.sciencecareers .org/. For more information about the history and content of myIDP, see Chapter 8.
5. Find a counselor near you through the National Board for Certified Counselors: http://www.nbcc.org/counselorFind.

2. But I Have No Skills! (and Other Myths)

1. Young 2011.
2. http://www.nationalpostdoc.org/?CoreCompetencies. See also Chapter 4 for a list of these competencies.
3. Information about myIDP can be found at https://myidp.sciencecareers .org/.
4. Michigan State University, n.d.
5. Adapted from Borchard, Bonner, and Musich 2002.

3. How to Identify Your Personal Values

1. Adapted from University of Denver Student Life Career Services, n.d.

4. To Postdoc or Not to Postdoc?

1. See, for example, Kaplan 2012.
2. The National Postdoctoral Association (NPA), a nonprofit organization focused on advocacy for postdoctoral scholars, defines "postdoctoral scholar" this way:

> A postdoctoral scholar ("postdoc") is an individual holding a doctoral degree who is engaged in a temporary period of mentored research and/or scholarly training for the purpose of acquiring the professional skills needed to pursue a career path of his or her choosing.

This more or less mirrors the definition proposed by the National Institutes of Health (NIH) and the National Science Foundation (NSF):

> An individual who has received a doctoral degree (or equivalent) and is engaged in a temporary and defined period of mentored advanced training to enhance the professional skills and research independence needed to pursue his or her chosen career path.

By contrast, the National Academies of Sciences defines the term "postdoc" this way:

- The appointee has received a PhD or doctorate equivalent.
- The appointment is viewed as an apprenticeship—a training or transitional period preparatory to a long-term academic, industrial, governmental, or other full-time research career.
- The appointment involves full-time research or scholarship.
- The appointment is temporary.
- The appointee is expected to publish (and receive credit for) the results of research or other activities performed during the period of the appointment.

The definition above was derived from the *Report and Recommendations* by the Committee on Postdoctoral Education of the Association of American Universities and others. While this word is defined in various ways, common threads found among the definition indicate that a postdoc requires a doctoral degree, involves full-time research, provides additional training, and is temporary.

Several definitions include the word "mentored," but that word is such a subjective descriptor that I hesitate to list it among common postdoc traits. Suffice it to say that postdoctoral training takes place under the supervision of a faculty member or PI.

Note that I also omit the phrase "needed to pursue his/her chosen career path." I do this by design, as a postdoc is *not* in fact required for many careers that a PhD in STEM might consider. These will be discussed in more detail in Chapter 5.

3. B., Sonja 2011.
4. Stephan 2014, p. 157.
5. For more information on transitional experiences, see Chapter 6.
6. National Science Foundation 2013a.
7. Ibid.
8. Bonetta 2008.
9. National Science Foundation 2013a.
10. See National Postdoctoral Association, n.d., "Postdoctoral Teaching Fellowships."
11. For more information, visit the Fellowships page at AAAS: http://www .aaas.org/page/fellowships.
12. For more information on the study, see Appendix A.
13. Ferguson et al. 2014.
14. For specific goal setting exercises and templates on drafting an IDP, see Chapter 8.
15. P. Clifford, personal communication, May 2014. The outline for developing an IDP first appeared on the FASEB website in July 2002 and was followed by a series of articles to explain the process. Philip Clifford, a scientist, university leader, and active member of FASEB, was instrumental in putting forward this initiative and then spreading the

word across the scientific community. In 2005, he led a workshop at the Experimental Biology annual meeting to explore the IDP process in more detail, and he has offered innumerable workshops, seminars, and authored papers on the topic since. He also serves as an author of myIDP, an online version of an IDP process first published in 2011, a worthwhile resource with automated reminders of goals set by the user, as well as helpful articles and descriptions of various fields. For more information, visit myIDP at http://myidp.sciencecareers.org/.

16. See the Sigma Xi Postdoc Survey findings at Davis 2005.
17. For more on the responsibilities of the postdoc and the mentor, see the Association of American Medical Colleges 2006.
18. See National Postdoctoral Association 2009.
19. National Academy of Sciences 2000.
20. For an excellent discussion of time-management strategies, see Kearns and Gardiner 2006a,b,c.

5. Career Options for PhDs in Science

1. National Science Foundation, "About the National Center for Science and Engineering Statistics," http://www.nsf.gov/statistics/about-ncses .cfm.
2. One of the surveys conducted via the NCSES is the Survey of Earned Doctorates (SED). This instrument collects information on an individual's educational background, demographic information, and postgraduate plans. The response rate of this instrument is 96 percent, incredibly high in survey research, but the chief limitation of this data set is the lack of information from PhD recipients who earned their degrees outside of the United States. Another limitation is the fact that this data source includes only plans made at the time of graduation, perhaps before a candidate has found a permanent job or completed additional postdoctoral training.

   Another data set is the Survey of Doctorate Recipients (SDR), a longitudinal study which follows a sample of respondents taken from the SED and tracked over time. The SDR includes information on demographics, educational history, employment status, and occupation. This data collection certainly represents a more accurate picture of currently employed PhDs in scientific disciplines, but again measures only those PhDs earned within the United States, as it is drawn from the SED.

   The NSF measures postdocs within the United States through its Survey on Graduate Students and Postdoctorates in Science and Engineering (GSS). This data set includes numbers of graduate students, postdocs, and doctorate-holding nonfaculty researchers. While this survey does indeed measure those postdocs whose PhD training took place outside of the United States, it does not include information on career plans of this group, nor does it capture postdocs in industry or nonprofit organizations.

In the fall of 2015, the NSF engaged in a full-scale launch of the Early Career Doctorates Survey (ECDS), an instrument designed to address some of the existing gaps in data mentioned previously. This survey will include information on respondents' educational history, professional activities, employer characteristics, work/life balance, mentorship, training, and research opportunities, and career paths and plans. Unfortunately, at the time of publication, the data set generated by the ECDS was not yet available.

3. Davis 2005.

4. Helm, Campa, and Moretto 2012.

5. After securing approval to conduct human subjects research by the Harvard University Committee on the Use of Human Subjects in 2015, a pilot phase was launched to test the instrument, and after revision, a full-scale launch of the survey took place in April 2015.

6. For an analysis of the skills developed during graduate school and those required for success in occupations, see Chapter 2.

7. For more on the "adjunctification" of the PhD workforce, see Jenkins 2014. In 2016 more than half of all faculty members in the United States hold part-time, contingent appointments.

8. National Science Foundation 2013b.

9. Lauren Celano, personal communication, February 13, 2016.

10. Freedman 2008.

11. See BEST, "About Best," http://www.nihbest.org/about-best/, and "17 Research Sites," http://www.nihbest.org/about-best/17-research-sites/.

6. Strategies for Exploring Careers and Building Experience

1. Adapted from a job description at http://careercenter.autm.net/jobseeker /job/21287887/Technology%20Transfer%20Associate/__company__ /?vnet=0, accessed January 16, 2015, no longer available.

2. Details of the Summer Associate program at the RAND Corporation may be found at http://www.rand.org/about/edu_op/fellowships/gsap .html, accessed January 21, 2015.

3. Details of the Presidential Management Fellowship Program may be found at http://www.pmf.gov/the-opportunity/pmf-stem.aspx, accessed January 21, 2015.

4. Minton 2013.

5. According to the survey research I conducted in 2015, 69 percent of all PhDs who are currently employed found their positions through networking.

7. How to Network Effectively

1. Laura Stark, PhD, personal communication, January 12, 2013.
2. Weinstock 2013.
3. Maia Weinstock, https://www.flickr.com/photos/pixbymaia/sets/721576 23988000684/.
4. Society for Human Resource Management 2013.
5. Society for Industrial and Applied Mathematics 2012.
6. Harvard University, Office of Career Services n.d.

8. How to Craft Your Individual Development Plan

1. Preston 2002.
2. Ibid.
3. Rieff 2002.
4. For more information, see Federation of American Societies for Experimental Biology, n.d.
5. See National Institutes of Health 2013.
6. Vanderbilt University School of Medicine n.d.
7. This exercise is adapted from work by Mary Todd, former director, Career Services, Bunker Hill Community College, Boston, MA.

9. How to Apply for Jobs

1. Adapted from "Susan Ireland's Resume Site," http://susanireland.com /resume/how-to-write/.
2. Krannich and Enelow 2002.
3. For information on drafting these application materials, see Vick, Furlong, and Lurie 2016.
4. PBS NewsHour 2012.
5. Find an American Job Center near you: http://www.careeronestop .org/localhelp/local-help.aspx.

10. How to Interview and Negotiate

1. Cosentino 2013.
2. Sample questions are adapted from Cobb, n.d.
3. For the American Association of University Professors (AAUP) Salary Survey, see http://www.aaup.org/reports-publications/2013-14salary survey; for the American Chemical Society Salary Survey, see http://cen .acs.org/articles/92/i35/2014-Salaries-Employment.html; and for the Annual Survey of the Mathematical Sciences, see http://www.ams .org/profession/data/annual-survey/docsgrtd.

Conclusion

1. Amy Rawls, personal communication, February 25, 2015.
2. Ibid.

Appendix A

1. The text in this appendix also appears in Sinche et al., ms. in prep.
2. Patton 1990.
3. Davis 2006; Sauermann and Roach 2012; Gibbs and Griffin 2013.
4. See Behr et al. 2012 for a discussion of using probing questions in web surveys.
5. Bogdan and Biklen 2006.
6. Sauermann and Roach 2013.

References

American Physiological Society. 2003. "APS/ACDP List of Professional Skills for Physiologists and Trainees." Available at http://www .lifescitrc.org/resource.cfm?submissionID=2625.

"Are 'Alternative' PhD Careers Still the Alternative?" LinkedIn group "Alternative PhD Careers," October 2013.

Association of American Medical Colleges. 2006. "Compact between Postdoctoral Appointees and Their Mentors." Drafted by AAMC Group on Graduate Research, Education, and Training (GREAT), and its Postdoctorate Committee. Available at https://www.aamc .org/initiatives/research/postdoccompact/.

Association of American Universities, Committee on Postdoctoral Education. 1998, March 31. "Report and Recommendations." Available at https://www.aau.edu/WorkArea/DownloadAsset.aspx?id=6834.

B., Sonja. 2011. "To Postdoc or Not to Postdoc? Musings from a Vancouver Scientist." *The Black Hole.* Accessed January 5, 2015. http://scienceadvocacy.org/Blog/2011/04/25/to-postdoc-or-not-to -postdoc/, no longer available.

Behr, Dorothée, Lars Kaczmirek, Wolfgang Bandilla, and Michael Braun. 2012. "Asking Probing Questions in Web Surveys: Which Factors Have an Impact on Quality of Reponses?" *Social Science Computer Review* 30(4): 487–498.

Bogdan, Robert C., and Sari K. Biklen. 2006. *Qualitative Research for Education: An Introduction to Theories and Methods,* 5th ed. Boston: Pearson.

Bonetta, Laura. 2008, June 13. "Industrial Postdocs: The Road Less Traveled." *Science,* Special Feature. Available at http://www .sciencemag.org/careers/features/2008/06/industrial-postdocs-road -less-traveled.

Borchard, David, C. Bonner, and S. Musich. 2002. *Your Career Planner,* 8th ed. Dubuque, IA: Kendall/Hunt Publishing.

Clifford, Phillip. 2002. "Quality Time with Your Mentor." *The Scientist* 16(19): 59. Available at http://www.the-scientist.com/?articles.view /articleNo/14274/title/Quality-Time-with-Your-Mentor/.

Cobb, Leslie. n.d. "Illegal or Inappropriate Interview Questions." Lawrence Berkeley National Laboratory, Berkeley CA. Available at http://www.gsworkplace.lbl.gov/DocumentArchive /BrownBagLunches/IllegalorInappropriateInterviewQuestions.pdf.

Cosentino, Marc. 2013. *Case in Point: Complete Case Interview Preparation,* 8th ed. Santa Barbara, CA: Burgee Press.

Creswell, John W. 2014. *Research Design: Qualitative, Quantitative, and Mixed Methods Approaches,*4th ed. Thousand Oaks, CA: Sage Publications.

Davis, Geoff. 2005. "Doctors Without Orders." *American Scientist* 93(3), special supplement. Available at http://www.sigmaxi.org/docs/default -source/Programs-Documents/Critical-Issues-in-Science/postdoc -survey/highlights.

———. 2006. "Improving the Postdoctoral Experience: An Empirical Approach." In *The Science and Engineering Workforce in the United States,* ed. R. Freeman and D. Goroff. Chicago: University of Chicago Press.

Federation of American Societies for Experimental Biology. n.d. "Individual Development Plan for Postdoctoral Fellows." Available at http://www.faseb.org/portals/2/pdfs/opa/idp.pdf.

Ferguson, Kryste, B. Huang, L. Beckman, and M. Sinche. 2014. "National Postdoctoral Association Institutional Policy Report 2014: Supporting and Developing Postdoctoral Scholars." Washington, DC: National Postdoctoral Association.

Fiske, Peter S. 2001. *Put Your Science to Work: The Take-Charge Career Guide for Scientists.* Washington, DC: American Geophysical Union.

Freedman, Toby. 2008. *Career Opportunities in Biotechnology and Drug Development.* Cold Spring Harbor, NY: Cold Spring Harbor Laboratory Press.

Freeman, Richard B., and D. L. Goroff, eds. 2009. *Science and Engineering Careers in the United States: An Analysis of Markets and Employment.* Chicago: University of Chicago Press.

Fuhrmann, Cynthia N., J. A. Hobin, B. Lindstaedt, and P. S. Clifford. 2011. "myIDP." American Association for the Advancement of Science. Available at http://myidp.sciencecareers.org/.

Gaensler, Bryan. 2015, March 3. "Maybe the Hardest Nut for a New Scientist to Crack: Finding a Job." "The Conversation" blog.

Available at https://theconversation.com/maybe-the-hardest-nut-for-a
-new-scientist-to-crack-finding-a-job-37492.

Gibbs, Kenneth D., Jr., and Kimberly A. Griffin. 2013. "What Do I Want
to Be with My PhD? The Roles of Personal Values and Structural
Dynamics in Shaping the Career Interests of Recent Biomedical
Science PhD Graduates." *CBE Life Sciences Education* 12(4):
711–723.

Gibbs, Kenneth D., Jr., John McGready, Jessica C. Bennett, and Kimberly A. Griffin. 2014. "Biomedical Science Ph.D. Career Interest
Patterns by Race/Ethnicity and Gender." *PLoS One* 9(12): s5.

Harvard University, Office of Career Services. n.d. "Building Professional
Connections." Available at http://ocs.fas.harvard.edu/files/ocs/files
/gsas-building-connections-publication.pdf.

Helm, Matt, Henry Campa III, and Kristin Moretto. 2012. "Professional
Socialization for the Ph.D.: An Exploration of Career and Professional Development Preparedness and Readiness for Ph.D. Candidates." *Journal of Faculty Development,* 26(2): 5–23.

Holland, John. 1997. *Making Vocational Choices: A Theory of Vocational
Personalities and Work Environments,* 3rd ed. Lutz, FL: Psychological Assessment Resources.

Iasevoli, Brenda. 2015, February 27. "A Glut of PhDs Means Long Odds
of Getting Jobs." nprEd blog, WBUR. Available at http://www.npr
.org/blogs/ed/2015/02/27/388443923/a-glut-of-ph-d-s-means-long
-odds-of-getting-jobs.

Janssen, Kaaren, and R. Sever, eds. 2015. *Career Options for Biomedical
Scientists.* Cold Spring Harbor, NY: Cold Spring Harbor Laboratory
Press.

Jenkins, Rob. 2014. "Straight Talk about 'Adjunctification.'" *The Chronicle of Higher Education,* December 15. Available at http://chronicle
.com/article/Straight-Talk-About/150881/.

June, Audrey W. 2014. "Doctoral Degrees Increased Last Year, but Career
Opportunities Remained Bleak." *The Chronicle of Higher Education,* December 19. Available at http://chronicle.com/article/
Doctoral-Degrees-Increased/150421/?cid=gs&utm_source=gs&utm
_medium=en.

Kaplan, Karen. 2012. "Postdoc or Not?" *Nature* 483: 499–500.
doi:10.1038/nj7390-499a.

Kearns, Hugh, and M. Gardiner. 2006a. *Defeating Self-Sabotage: Getting
Your PhD Finished.* Sydney, Australia: Thinkwell.

———. 2006b. *The Seven Secrets of Highly Effective PhD Students.*
Sydney, Australia: Thinkwell.

———. 2006c. *Time for Research: Time Management for Academics,
Researchers, and PhD Students.* Sydney, Australia: Thinkwell.

Krannich, Ronald L., and W. S. Enelow. 2002. *Best Resumes and CVs for International Jobs: Your Passport to the Global Job Market.* Manassas Park, VA: Impact Publications.

Lametti, Daniel. 2012. "Is a Science Ph.D. a Waste of Time?" *Slate.* Available at http://www.slate.com/articles/health_and_science/science /2012/08/what_is_the_value_of_a_science_phd_is_graduate_school _worth_the_effort_.html.

Michigan State University. n.d. "Career Success: Starting Your Self-Assessment." Available at http://careersuccess.msu.edu/assessments.

Minton, Peter I. 2013. "6 Legal Requirements for Unpaid Internship Programs." *Forbes,* April 19. Available at http://www.forbes.com/sites /theyec/2013/04/19/6-legal-requirements-for-unpaid-internship -programs/.

National Academy of Sciences. 1995. *Reshaping the Graduate Education of Scientists and Engineers.* Washington, DC: National Academies Press.

——. 2000. *Enhancing the Postdoctoral Experience for Scientists and Engineers.* Washington, DC: National Academies Press.

——. 2007. *Rising above the Gathering Storm: Energizing and Employing America for a Brighter Economic Future.* Washington, DC: National Academies Press.

——. 2014. *The Postdoctoral Experience Revisited.* Washington, DC: National Academies Press.

National Institutes of Health. 2013, July 23. "NIH Encourages Institutions to Develop Individual Development Plans for Graduate Students and Postdoctoral Researchers." Available at https://grants .nih.gov/grants/guide/notice-files/NOT-OD-13-093.html.

——. 2014, August 4. "Revised Policy: Descriptions on the Use of Individual Development Plans (IDPs) for Graduate Students and Postdoctoral Researchers Required in Annual Progress Reports beginning October 1, 2014." Available at https://grants.nih.gov/grants /guide/notice-files/NOT-OD-14-113.html.

National Postdoctoral Association. n.d. "What Is a Postdoc?" Accessed December 2, 2014. http://www.nationalpostdoc.org/?page=What _is_a_postdoc.

——. n.d. "Postdoctoral Teaching Fellowships." http://www.national postdoc.org/?page=TeachingFellowships.

——. 2009. "NPA Core Competencies." http://www.nationalpostdoc .org/?CoreCompetencies.

National Research Council. 1998. *Trends in the Early Careers of Life Scientists.* Committee on Dimensions, Causes, and Implications of Recent Trends in the Careers of Life Scientists. Washington, DC: National Academies Press.

———. 2005. *Bridges to Independence: Fostering the Independence of New Investigators in Biomedical Research.* Washington, DC: National Academies Press.

National Science Foundation. Survey of Earned Doctorates. 2000–2009. http://www.nsf.gov/statistics/srvydoctorates.

———. 2008. Survey of Doctorate Recipients. http://www.nsf.gov /statistics/srvydoctoratework/.

———. 2013a. Survey of Doctorate Recipients. Table 77. "Doctoral Scientists and Engineers Employed as Postdoctoral Appointees, by Selected Demographic Characteristics and Broad Field of Doctorate: 2013." http://www.nsf.gov/statistics/srvydoctoratework.

———. 2013b. Survey of Doctorate Recipients, Special Tabulation. S. Proudfoot & D. Foley.

———. 2013c. Survey of Graduate Students and Postdoctorates in Science and Engineering. http://www.nsf.gov/statistics /srvygradpostdoc/.

Oxford Psychologists Press. 2015. "Myers' and Briggs' Famous Tool Is Still Going Strong, Constantly Being Improved." February 4. Available at https://www.opp.com/en/tools/MBTI/Myers-Briggs -history.

Patel, Vimal. 2015. "Pushing for Culture Change, Ph.D.'s Explore Careers Beyond Academe." *The Chronicle of Higher Education,* March 20.

Patton, Michael Q. 1990. *Qualitative Evaluation and Research Methods,* 2nd ed. Newbury Park, CA: Sage Publications.

Patton, Stacey. 2014, July 30. "Between Postdoc and Job, a Whole Lot of Questions." *Vitae.* Available at https://chroniclevitae.com/news/632 -between-postdoc-and-job-a-whole-lot-of-questions?cid=oh&utm _source=oh&utm_medium=en.

PBS Newshour, 2012, September 25. "Is Applying for Jobs Online Not an Effective Way to Find Work?" Available at http://www.pbs .org/newshour/bb/business-july-dec12-makingsense_09-25/.

Preston, Julian. 2002. "Mentors Are Made, Not Born." *The Scientist,* September 2. Available at http://www.the-scientist.com/?articles.view /articleNo/14218/title/Mentors-are-Made—not-Born/.

Rieff, Heather. 2002. "A Mentor Maintenance Program." *The Scientist,* October 14. Available at http://www.the-scientist.com/?articles.view /articleNo/14306/title/A-Mentor-Maintenance-Program/.

Rosen, Stephen, and Celia Paul. 1998. *Career Renewal.* San Diego, CA: Academic Press.

Sauermann, Henry, and M. Roach. 2012. "Science PhD Career Preferences: Levels, Changes, and Advisor Encouragement." *PLoS One* 7(5): e36307.

———. 2013. "Increasing Web Survey Response Rates in Innovation Research: An Experimental Study of Static and Dynamic Contact Design Features." *Research Policy* 42(1): 273–286.

Schillebeeckx, Maximiliaan, Brett Maricque, and Cory Lewis. 2013. "The Missing Piece to Changing the University Culture." *Nature Biotechnology* 31(10): 938.

Selfa, Lance A., and Steven Proudfoot. 2014. "Unemployment among Doctoral Scientists and Engineers Remained Below the National Average in 2013." NCSES InfoBrief, National Science Foundation, September. Available at http://www.nsf.gov/statistics/infbrief /nsf14317/.

Sher, Barbara, with Barbara Smith. 1994. *I Could Do Anything If I Only Knew What It Was*. New York: Dell.

Sinche, Melanie,* R. L. Layton,* P. D. Brandt, A. M. Freeman, J. R. Harrell, and J. D. Hall. ms. in prep. "An Evidence-Based Evaluation of Transferable Skills Acquired during Doctoral Training in STEM." * indicates equal credit

Society for Human Resource Management. 2013, April 11. "SHRM Survey Findings: Social Networking Websites and Recruiting/ Selection." Available at http://www.shrm.org/research/surveyfindings /articles/pages/shrm-social-networking-websites-recruiting-job -candidates.aspx.

Society for Industrial and Applied Mathematics. 2012. *Mathematics in Industry*. Accessed Feb. 17, 2015. http://www.siam.org/reports/mii /2012/index.php.

Stephan, Paula. 2014. *How Economics Shapes Science*. Cambridge, MA: Harvard University Press.

Teitelbaum, Michael S. 2014a. *Falling Behind? Boom, Bust, and the Global Race for Scientific Talent*. Princeton, NJ: Princeton University Press.

———. 2014b. "The Myth of the Science and Engineering Shortage." *The Atlantic*, March 19. Available at http://www.theatlantic.com /education/archive/2014/03/the-myth-of-the-science-and-engineering -shortage/284359/.

University of Denver Student Life Career Services. n.d. "Work Values." Available at http://www.du.edu/career/media/documents/pdfs /workvalues.pdf.

Vanderbilt University School of Medicine, Office of Biomedical Research and Training, Office of Postdoctoral Affairs. n.d. "Individual Development Plan for Postdoctoral Fellows at Vanderbilt University." Available at https://medschool.vanderbilt.edu/postdoc/individual -development-plan-idp-postdoctoral-fellows-vanderbilt-university.

Vastag, Brian. 2012. "U.S. Pushes for More Scientists, but the Jobs Aren't There." *The Washington Post,* July 7. Available at https://www
.washingtonpost.com/national/health-science/us-pushes-for-more
-scientists-but-the-jobs-arent-there/2012/07/07/gJQAZJpQUW_story
.html.

Vick, Julia Miller, Jennifer S. Furlong, and Rosanne Lurie. 2016. *The Academic Job Search Handbook,* 5th ed. Philadelphia: University of Pennsylvania Press.

Weinstock, Maia. 2013. "Breaking Brick Stereotypes: LEGO Unveils a Female Scientist." *Scientific American,* September 2. Available at http://blogs.scientificamerican.com/guest-blog/breaking-brick
-stereotypes-lego-unveils-a-female-scientist/.

Weissmann, Jordan. 2013. "The Ph.D. Bust: America's Awful Market for Young Scientists—in 7 Charts." *The Atlantic,* February 20. Available at http://www.theatlantic.com/business/archive/2013/02/the-phd-bust
-americas-awful-market-for-young-scientists-in-7-charts/273339/.

Young, Valerie. 2011. *The Secret Thoughts of Successful Women: Why Capable People Suffer from the Impostor Syndrome and How to Thrive in Spite of It.* New York: Crown Business.

Acknowledgments

For all scientists, I feel compelled to share the stories of this book, illustrating that there are countless scientists who enjoy their work, and to share the knowledge and experience I have acquired over time to assist scientists with their own discoveries.

The community of professionals who study careers for PhDs in science has increased exponentially since I entered the field in 1998, and I have grown as a professional through my interactions with these peers—in particular, with members of the Graduate Career Consortium and the National Postdoctoral Association. For our lively and informative discussions, meetings, conferences, phone conversations, and email exchanges, I am thankful.

I am likewise grateful to the Labor and Worklife Program at Harvard Law School for supporting me throughout my research study connected to this book. In particular, I thank Richard Freeman, Elaine Bernard, and Lorette Baptiste for their ongoing support.

I need also to acknowledge experts who served as advisors to me during this process, including Lauren Celano, James Cuff, Geoff Davis, Jennifer Dineen, Peter Einaudi, Daniel Foley, Deneen Hatmaker, Becky Holmes-Farley, Hugh Kearns, Bill Kolata, Victoria McGovern, Sharon Milgram, Kelly Phou, Steven Proudfoot, Emilda Rivers, Michael Roach, Fadil Santosa, Henry Sauermann, Paula Stephan, Michael Teitelbaum, and Gordon Willis.

Thanks to the scientists who volunteered their time to assist me by interviewing for this book, including Steve Bennett, Holly Dail, Raluca Ellis, Richard Goldberg, Jim Gould, Rohan Manohar, Leslie Pond, Amy Rawls, Manisha Sinha, and Alok Tayi, as well as to those scientists who offered their job search documents.

I am thankful for those who supported me throughout this endeavor, offering either emotional or editorial support or both, especially Andrew Green and Laura Stark, for being with me in the very beginning and at the very end, and including Joseph Barber, Phil Clifford, Kryste Ferguson, Michelle Ferguson, Garth Fowler, Kenny Gibbs, Jim Gould, Belinda Huang, Bill Lindstaedt, Annie Marcucci, Keith Micoli, Kathy Mitchell, Walt Nakonechny, Michaela Tally, and Jill Voll.

Heartfelt thanks goes to Rebekah Layton and her team at the University of North Carolina at Chapel Hill for interest in my project and for careful analysis of the data I collected on PhDs in science.

I am grateful to the Jackson Laboratory for its ongoing support of my work, and in particular to Tom Litwin.

Some themes in Chapter 7 grew out of a conference workshop given by the author with Laura Stark, PhD, on February 15, 2013 at the American Association for the Advancement of Science Annual Meeting.

For the editorial support of Andrew Kinney, Katrina Vassallo, Kate Brick, and others at Harvard University Press, I am likewise thankful.

Thank you to my parents, Dr. Jerome and Lorraine Vigil, and to my in-laws, Charlie and Sheryl Sinche, for their belief in my work over the years.

For their constant love, support, and inquiries about word counts, I thank Charles and Henry, and for his critical guidance and steadfast love, Bryan Sinche.

Finally, thank you to the scientists I have met over the years who have shared their stories with me. You have enriched my life, and for that, I will always be grateful.

Index

Academe, vs. working world, 215–218

Academic jobs: "adjunctification," 83; alternatives to, 2–5; career satisfaction in, 81–83; culture favoring, 1; interviews for, 195–196, 204, 207–209; negotiating offer, 210–214; new positions vs. new PhDs, 1, 2; PhD, 84–87, 231; postdoc appointments, 56; postdoc required for, 49, 51–53; resources on, 231

"Adjunctification," 83

Alumni, networking with, 133–135

American Association for the Advancement of Science (AAAS) Fellowships, 59, 229

American Association of University Professors (AAUP) Salary Survey, 212

American Chemical Society (ACS), 72

American Chemical Society Salary Survey, 212

American Job Centers, 190

Annual Survey of the Mathematical Sciences, 212

Anxiety, 10

Association for Women in Science (AWIS), 72

Association of University Technology Managers (AUTM), 110, 231

Barriers, to career development, 155–157

Behavior-based questions, in interviews, 199–200, 207

Bennett, Steve, 101–102, 121

BEST (Broadening Experiences in Scientific Training), 97

Best Resumes and CVs for International Jobs: Your Passport to the Global Job Market (Krannich and Enelow), 180

Biogen Idec, 104

Bioscience courses, 234

Biotechnology, 89–91

Blogging, 132

Board members, contacting, 111

Boston Business Journal, 205

Broadening Experiences in Scientific Training (BEST), 97

Business: MBA earnings vs. PhD, 50–51; mini-MBA, 99, 232; resources on, 232

Business cards, 116, 188, 209

Career Beliefs Inventory (CBI), 17–18
Career counselors: feedback from,
14–15; finding, 19; formal assessments by, 17–19; job search with,
189–190; networking help from,
134–135
Career development: barriers to,
155–156; continual cycle of, 105;
Individual Development Plan,
67–69, 145–157; postdoc programs, 70–71; stages of, 10–12
Career development profile,
45–46
Career exploration, 11, 107–123;
gaining experience in, 107, 116–122;
identifying careers of interest in,
123; informational interviews for,
111–116; internship or fellowship
for, 119–120; networking for, 122;
professional association resources
for, 110–111, 112; project collaboration for, 116–118; reading
job descriptions, 107–110;
resources for, 230; shadowing for,
121–122; taking classes for,
120–121
Career fairs, 187–188
*Career Opportunities in Biotech-
nology and Drug Development*
(Freedman), 91
Career options for PhDs. *See* PhD
careers
Career satisfaction, 3, 80–83;
measurement of, 81–83; in
tenure-track vs. non-tenure-track
jobs, 81–83; transition and, 220
Case in Point (Cosentino), 197
Case interviews, 196–197
Category headings, on resume,
176–177
CBI. *See* Career Beliefs Inventory
Celano, Lauren, 90–91
Chance encounters, 190
Children, postdoc and, 74–75
Chronicle of Higher Education, 62,
87, 207

Classes, career exploration through,
120–121
Clifford, Philip, 147–148
Climate and Urban Systems Partnership (CUSP), 97–99, 117–118
Climate Corporation, The, 100–101
Clontech, 102
CNET, 205
Coding skills, 99–100, 101, 121
Collaboration: on IDP, 146–147;
on project, 116–118
Community college jobs, 85–86
Comprehensive/regional university
jobs, 85–86
Consulting: case interviews for,
196–197; internships, 119; job
satisfaction in, 82; offering services
in, 118; PhDs engaged in, 92, 95;
sample cover letter for, 183; sample
resume for, 178–180; self-made
postdoc in, 60; transferable skills
for, 24–25; values and, 45
Consulting club, 70–71, 199
Core competencies, 23, 69–70
Cosentino, Marc, 197
Courtesy, in interview, 194
Cover letter, 180–181; exercise on
writing, 180; Q&A on, 181, 184;
sample, for consulting position,
183; sample, for research position,
182–183
Culture, organizational, 216–217
"Culture fit," 217
Curriculum Vitae. *See* CV
CV, 159–180; conversion to resume,
163–180; drafting, 160–163; formatting tips, 162; highlighting
transferable skills on, 32–35; international vs. U.S., 177–180; interview
questions about, 198; resume vs.,
159–160; sample, 164–169

Dail, Holly, 100–101
Data science jobs: resources on,
112, 234; sample resume for,
172–174

Death/grief, dealing with during postdoc, 77–78
Dress, professional, 194, 217
Drug development, 89–91, 104

Economic barriers, 156
Educational barriers, 156
Education contacts, networking via, 133–135
Education jobs, for PhDs, 84–87, 231. *See also* Faculty positions
Elementary education, PhD jobs in, 85–86
Elevator speech, 126–127
Ellis, Raluca, 97–99, 117–118
Email: accepting position via, 66; contacting board members via, 111; job news via, 209–210; myIDP, 19; networking via, 128, 130, 132, 134–139, 142; postdoc application via, 61–62; requesting informational interview via, 113, 114; thank you for interview via, 116, 209
Enthusiasm, in interview, 194–195
Entrepreneurship: PhD careers, 99–100; self-made internships, 119–120; self-made postdocs, 60
Environment: college campus, 87; identifying preferred, 16
Experience, gaining, 107, 116–122; though collaboration on project, 116–118; though internship or fellowship, 119–120; through shadowing, 121–122; through taking classes, 120–121; through volunteering, 116, 117–118

Facebook, 131
Faculty advisor: IDP collaboration with, 146–147; IDP sharing with, 153–155; postdoc application to, 61–62; postdoc goal discussion with, 67–69; postdoc information

(letter) from, 65–66; postdoc mentorship by, 72–73; postdoc questions for, 64–65; postdoc support lacking from, 75–76; project collaboration with, 116–118
Faculty positions: "adjunctification," 83; alternatives to, 2–5; career satisfaction in, 81–83; culture favoring, 1; CV for, 159–180; interviews for, 195–196, 204, 207–209; negotiating offer, 210–214; new positions vs. new PhDs, 1, 2; PhD jobs, 84–87, 231; postdoc appointments, 56; postdoc required for, 49, 51–53; resources on, 231
Family: networking via, 132–133; postdoc and, 74–75, 77–78
Fast Company, 205
Federation of American Societies of Experimental Biology (FASEB), 68, 145–147
Feedback: from career counselors and mentors, 14–15; from others, 13–14
Fellowships, 119–120. *See also* Postdoc(s)
Field-specific postdocs, 59–60
Finances: economic barriers to career development, 156; economics of postdoc, 50–51; elimination of postdoc funding, 77; mean salaries by education, 51, 52; negotiating offer, 210–214
Fit (good fit), 4–5, 44–45, 107, 217, 220
Formal assessments, 17–19
Franklin Institute Science Museum, 97–99, 117–118
Freedman, Toby, 91
Friends, networking via, 132–133
Fuhrmann, Cynthia, 147–148
Future: acknowledging thoughts about, 10

Goal setting: career, 11, 67; Individual Development Plan, 149–153; postdoc, 67–69; transition to work environment, 218
Good fit, 4–5, 44–45, 107, 217, 220
Gould, Jim, 103–104
Government jobs, for PhDs, 84, 87–89, 231
Government postdocs, 57–59
Graduate Career Consortium, 225
Grant administrators, networking with, 135–136
Grief, over loss during postdoc, 77–78

Harvard Medical School, 103–104
Headhunters, 188
Hierarchies, organizational, 216
Hobin, Jennifer, 147–148
How Economics Shapes Science (Stephan), 50–51

Ideal job: drafting description, 16; finding, 15–16
Identifying Career Pathways for PhDs in Science (Harvard University), 80–84, 223–228
Identity crisis, 215
IDP. *See* Individual Development Plan
Illegal questions, in interview, 200–203
Impostor syndrome, 21–22
Individual Development Plan (IDP), 67–69, 145–157; collaboration with advisor on, 146–147; components of, 147; exercises on, 149, 152–153; institutional use of, 148–149; introduction of, 68; myIDP, 19, 23, 147–148, 229; required by federal agencies, 148; reviewing and revising, 155; sharing with advisor, 153–155; SMART goals for, 149–153
Industry jobs, for PhDs, 84, 89–91, 232

Industry postdocs, 56–57, 58
Informational interviews, 111–116; definition of, 111; identifying professionals to interview, 114–115; networking vs., 139–140; questions to ask in, 115; requesting, 112–114
In-person interview, 194–195
Interaction with science, 15–16
Interests: exercises on, 13–16; feedback from career counselors and mentors on, 14–15; feedback from others on, 13–14; finding ideal job, 15–16; formal assessments, 17–19; identifying, 3, 9–19; interaction with science in job, 15–16; leisure, linking to jobs, 17
International CVs and resumes, 177–180
International scholars, postdocs for, 57
Internships, 119–120, 189
Interviews, informational, 111–116, 139–140
Interviews, job, 191–210; academic, 195–196, 204, 207–209; behavior-based questions in, 199–200, 207; blunders in, 191; case, 196–197; closing and following up, 209–210; courtesy in, 194; dos and don'ts, 211; enthusiasm in, 194–195; illegal questions in, 200–203; in-person, 194–195; nonverbal signals in, 195; practicing, 205–206; preparing for, 5–6, 203–209; professional dress for, 194; questions about CV or resume in, 198; questions about employer in, 203; question types in, 197–203; sample questions for, 206–209; Skype or other video, 193; standard questions in, 198; telephone, 192–193; types of, 192–197
Interviews, postdoc, 62, 63–65

Job descriptions, 107–110
Job search, 5–6; career counselor for, 189–190; career fairs in, 187–188; chance encounters and, 190; as final stage of career development, 10–11; Individual Development Plan for, 145–157; interviewing, 5–6, 191–210, 211; materials for, 5, 159–185; methods of, 185–190; negotiating offer, 212–216; networking in, 185–187; online applications, 185–187; placement firm or headhunter in, 188
Job titles, 85, 109
Journal databases, 136
Journal editors, networking with, 135
Journal postings, for postdocs, 62–63

Krumboltz, John, 17–18

Liberal arts colleges, PhD jobs at, 85–86
Lindstaedt, Bill, 147–148
LinkedIn, 128–131
Linn, Steve, 146

Manohar, Rohan, 102
Mathematics in Industry (Society for Industrial and Applied Mathematics), 137–138
MBA: abridged course, 99, 232; earnings vs. PhD, 50–51
MBTI. *See* Myers–Briggs Type Indicator
Medical device development, 89–91
Medical school jobs, 85–86, 103–104
Mentors: becoming, 220; changing, for transitional postdoc, 60; collaboration on IDP, 146–147; feedback from, 14–15; multiple, during postdoc, 72–73; peers as, 14–15, 70–71; seeking, in transition to work, 219; sharing IDP with, 153–155

Michigan State University Career Success program, 23
Mini-MBA, 99, 232
Minoritypostdoc.org, 63
Museum jobs, 91–93, 97–99, 117–118
Myers–Briggs Type Indicator (MBTI), 18
myIDP, 19, 23, 147–148, 229

National Academies of Science, 2, 72
National Association of Science Writers, 110
National Board for Certified Counselors, 19
National Cancer Institute, 60
National Center for Science and Engineering Statistics (NCSES), 79
National Institutes of Health, 57, 97
National Postdoctoral Association (NPA): core competencies, 23, 69–70; joining, 71; study of postdoc offices, 66; survey participation, 225; teaching fellowship list, 59
National Science Foundation (NSF): employment data, 79–84; IDP requirement, 148; postdoc data, 56, 57
National Science Teachers Association, 110
Nature, 62
Negotiation, 210–214
Networking, 5, 125–142; approaching contacts, 136–139; career exploration through, 122; chance encounters and, 190; definition of, 125; elevator speech, 126–127; feelings about, 5, 125; informational interview vs., 139–140; job search, 185–190; maintaining relationships over time, 142; mutual benefits of, 141–142; online, 128–132; online

Networking *(continued)*
job applications vs., 185–187;
stealth, 127; types of, 126–132;
using existing network for, 132–136
Nonprofit organizations, PhD jobs
with, 84, 91–93, 97–99, 231
Nonverbal signals, 195
Novartis Institutes for BioMedical
Research (NIBR), 57

One Stop Career Centers, 190
Online courses, 232–233
Online job applications, 185–187
Online networking, 128–132
"On Think Tanks: Independent
Research Ideas and Advice"
(onthinktanks.org), 205
Organizational culture, 216–217

Parent, loss during postdoc, 77–78
Peer mentorship, 14–15, 70–71
People, as factor in ideal job, 16
Personality: Myers–Briggs Type
Indicator, 18
Personal values. *See* Values
Petsko, Gregory, 1–2
Pharmaceuticals, 89–91, 104
PhD careers, 2–5, 79–105; academic,
vs. new PhDs, 1, 2; across all
sectors, 84–85; data/surveys on,
79–84, 223–228; demographics of
survey on, 81; dynamic possibilities, 105; education jobs, 84–87,
231; employment, five years out,
81; experience required for, 94–97;
government jobs, 84, 87–89, 231;
industry jobs, 84, 89–91, 232; job
titles, 85; need for information on,
2; nonprofit organization jobs, 84,
91–93, 97–99, 231; resources on,
229–233; snapshots of currently
employed PhDs, 97–104; transition
to working world, 6, 215–220;
work/activities in, 93–94, 95. *See
also* Career development; Career
exploration

Physical barriers, 156
PI. *See* Principal investigator
Placement firm, 188
Postdoc(s), 48–78; academic, 56;
applying to faculty member or PI
for, 61–62; applying to journal
postings for, 62–63; applying to
specific training programs for, 63;
building skills and experiences
during, 69–70; career and professional development during,
70–71; vs. career goal, 50;
challenges during, 75–78; decision
on pursuing, 49–50; description
for job requiring, 53–54; elimination of funding for, 77; employment survey on, 79–80; family/
child issues during, 74–75; fields
not requiring, 55; field-specific,
59–60; financial considerations in,
50–51; government, 57–59; grief
and loss during, 77–78; industry,
56–57, 58; interviewing for, 62,
63–65; lack of support from PI
during, 75–76; making most of,
65–75; multiple mentors during,
72–73; participation in professional groups during, 71–72;
positions requiring or preferring,
49, 50–54; questions to ask
faculty advisor or principal
investigator about, 64–65; research
and career goals in, 67–69;
searching for, 61–65; self-care
during, 73–74; self-made, 60;
successful, 78; teaching, 59;
termination without cause, 77;
transitional, 60; types of appointments, 56–60; for underrepresented groups, 63; written
information about, 65–66
Postdoc office, contacting, 66
*Postdoctoral Experience Revisited,
The,* 2
Postdoctoral training. *See* Postdoc(s)
Practice interviewing, 205–206

Pregnancy, during postdoc, 74–75
PreScouter, 99–100
Presidential Management Fellowship (PMF), 119
Preston, Julian, 146
Principal investigator (PI): IDP collaboration with, 146–147; IDP sharing with, 153–155; postdoc application to, 61–62; postdoc goal discussion with, 67–69; postdoc information (letter) from, 65–66; postdoc mentorship by, 72–73; postdoc questions for, 64–65; postdoc support lacking from, 75–76; project collaboration with, 116–118
Professional association: career exploration through, 110–111, 112; contacting board members of, 111; local meetings of, 110–111; networking via, 136; postdoc participation in, 71–72; websites of, 110–111
Propel Careers, 90–91
PsychInfo, 136
Psychological barriers, 156
PubMed, 136

Qualifications, summary of, 170–176
Questions, interview, 197–203; behavior-based, 199–200, 207; about CV or resume, 198; about employer, 203; illegal, 200–203; sample, 206–209; standard, 198

RAND Corporation, 119, 229
Rawls, Amy, 217
Reference letters, 184–185
Regulatory Affairs Professional Society, 110
Research Square, 217
Research university jobs, 85–86
Resources: for PhDs, 229–233; professional associations, 110–111, 112

Resume: category headings on, 176–177; CV conversion to, 163–180; CV vs., 159–160; exercise on writing, 170–171; international vs. U.S., 177–180; interview questions about, 198; relevant skills on, 171; sample, for consulting job, 178–180; sample, for data science job, 172–174; Summary of Qualifications on, 170–176
Rovi Corporation, 101–102

Salary.com, 211–212
Salary negotiations, 210–214
Salary surveys, 212
Sample cover letters, 182–183
Sample CV, 164–169
Sample resumes, 172–174, 178–180
Science, 62
Science writing, 110, 235
Scientific America, 127
Secondary education, PhD jobs in, 85–86
Secret Thoughts of Successful Women, The (Young), 21
Self-assessment: acknowledging thoughts about future, 10; drafting career development profile, 45–46; as first stage of career development, 11–12; formal tools for, 17–19; identifying interests, 3, 9–19; identifying skills, 3–4, 21–38; identifying values, 4, 39–45; Individual Development Plan, 149; need for, 9; risks in, 12
Self-care, during postdoc, 73–74
Self-made internships, 119–120
Self-made postdocs, 60
Self-reflection, 9, 13
SF Biotech, 205
Shadowing, 121–122
Shapiro, Julie, 169–174
Sigma Xi Postdoc Survey, 79–80
Sinha, Manisha, 104
Situation-actions-results, 198–200

Skills: broader application of, 22;
building, resources for, 230–231;
building by taking classes,
120–121; building during postdoc,
69–70; categories, on resumes,
176–177; developed by PhDs,
23–24; doubts about, 21–22;
exercises on, 24–26, 35–38;
identifying, 3–4, 21–38; myths
about, 21–23; NPA list of core
competencies, 23, 69–70; relevant,
on resume, 171; required for PhD
jobs, 94–97; transferable, deter-
mining and highlighting on CV, 24,
32; transferable, listing of, 230
Skype, 137–138, 193
SMART goals, 149–153
Social barriers, 156–157
Social media, 128–132
Society for Industrial and Applied
Mathematics, 137–138
Society for Neuroscience (SFN), 72
Society for the Advancement of
Chicanos and Native Americans
in Science (SACNAS), 72
Stark, Laura, 127, 129–130
Stealth networking, 127
Stephan, Paula, 50–51
Strong, E. K., Jr., 18
Strong Interest Inventory, 18
Successful postdoc, 78
Summary of Qualifications, 170–176

Tayi, Alok, 99–100
Teaching postdocs, 59
Teaching resources, 233
Teamwork, 216–217, 219
Technology transfer associate,
107–110
Technology Transfer Center, 60
Telephone interviews, 192–193
Temporary jobs (temping), 188–189
Tenure-track positions, 84–87;
alternatives to, 2–5; career

satisfaction in, 81–83; new
positions vs. new PhDs, 1, 2;
postdoc required for, 49, 51–53
Termination without cause, postdoc,
77
Thank-you note, for interview, 116,
209
Time requirements, in working
world, 216–217
Transferable skills: determining,
24–25; highlighting on CV, 32–35;
listing of, 233
Transition, to work environment, 6,
215–220; cultural differences in,
216–217; grace in, 218–220
Transitional postdocs, 60
Twitter, 131–132

Underrepresented groups: postdocs
for, 63; professional organizations
for, 72
Unemployment, 2–3
University of Hartford, 155

Values: exercise for clarifying, 40–44;
identifying, 4, 39–45; using to
determining fit, 44–45
Vanderbilt University School of
Medicine, 148–149
Video interviews, 193

Weinstock, Maia, 127
Woods Hole Oceanographic
Institution, 100–101
Work environment: cultural
differences in, 216–217;
grace in, 218–220; transition
to, 6, 215–220
Working scientists, work of, 93–94,
95
"Writing Experience," on resume,
176–177

Young, Valerie, 21